Human Factors
in Computer Systems

HUMAN/COMPUTER INTERACTION
A Series of Monographs, Edited Volumes, and Texts

SERIES EDITOR
BEN SHNEIDERMAN

Directions in Human/Computer Interaction
Edited by Albert Badre and Ben Shneiderman

Online Communities:
A Case Study of the Office of the Future
Starr Roxanne Hiltz

Human Factors In Computer Systems
Edited by John Thomas and Michael Schneider

Human Factors and Interactive Computer Systems
Edited by Yannis Vassiliou

Human Factors
in Computer Systems

Edited by

JOHN C. THOMAS
IBM T. J. Watson Research Center

MICHAEL L. SCHNEIDER
ITT Advanced Technology Center

ABLEX PUBLISHING CORPORATION
Norwood, New Jersey   07648

Cover. *Computer-drawn faces representing speech synthesizer controlling data. Computer graphics has become increasingly useful in the representation and interpretation of multidimensional data with complex relationships. One particularly novel method of representing multivariate data is a computer-drawn face. In general,* n *data parameters are each mapped into a figure with* n *features, each feature varying in size or shape according to the point's coordinate in that dimension. Faces allow the human analyst to visually grasp many of the essential regularities or irregularities in the data. Dr. Clifford Pickover of the IBM Watson Research Center has used such figures to represent the numerous data which controls an IBM speech synthesizer. Dr. Pickover has recently extended the use of computer-drawn faces to the field of molecular genetics by using them to characterize the statistical properties of the base sequence of a DNA cancer gene.*

Copyright © 1984 by Ablex Publishing Corporation

Second printing 1986

Library of Congress Cataloging in Publication Data
Main entry under title:

Human factors in computer systems.

   (Human/computer interaction)
   Bibliography: p.
   Includes indexes.
   1. Electronic data processing—Psychological aspects—
Addresses, essays, lectures.   2. Electronic digital
computers—Programming—Psychological aspects—Addresses,
essays, lectures.   I. Thomas, John, 1945 May 25–
II. Schneider, Michael, 1944 Oct. 3–   III. Series.
QA76.9.P75H86 1984   001.64   84-6371
ISBN 0-89391-146-1

Ablex Publishing Corporation
355 Chestnut Street
Norwood, New Jersey   07648

Contents

1

Heuristics for Designing Enjoyable User Interfaces:
Lessons from Computer Games

THOMAS W. MALONE

2

Learning to Use a Word Processor:
By Doing, By Thinking, and By Knowing

JOHN M. CARROLL AND ROBERT L. MACK

3

Formal Grammar as a Tool for Analyzing Ease of Use:
Some Fundamental Concepts

PHYLLIS REISNER

Contributors

ELIZABETH KRUESI BAILEY, *Software Metrics, Inc., Falls Church, Virginia 22046*

JOHN W. BAILEY, *Computer Metrics, Inc., Falls Church, Virginia 22046*

JOHN M. CARROLL, *IBM T.J. Watson Research Center, Yorktown Heights, New York 10598*

S. T. DUMAIS, *Bell Laboratories, Murray Hill, New Jersey 07974*

KATE EHRLICH, *Honeywell Information Systems, Waltham, Massachusetts 02154*

G. W. FURNAS, *Bell Laboratories, Murray Hill, New Jersey 07974*

TIMOTHY E. GOLDSMITH, *Psychology Department, New Mexico State University, Las Cruces, New Mexico 88003*

L. M. GOMEZ, *Bell Laboratories, Murray Hill, New Jersey 07974*

JAMES K. HABINEK, *IBM Development Laboratory, Rochester, Minnesota 55901*

T. K. LANDAUER, *Bell Laboratories, Murray Hill, New Jersey 07974*

ROBERT L. MACK, *IBM T. J. Watson Research Center, Yorktown Heights, New York 10598*

KENNETH MAGEL, *North Dakota University, Fargo, North Dakota 58105*

DANIEL G. McNICHOLL, *Department of Computer Science, University of Missouri, Columbia, Missouri*

THOMAS MALONE, *Sloan School of Management, Massachusetts Institute of Technology, Cambridge, Massachusetts 02139*

PHYLLIS REISNER, *IBM San Jose Research Center, San Jose, California 95193*

RICKY E. SAVAGE, *IBM Development Laboratory, Rochester, Minnesota 55901*

ROGER W. SCHVANEVELDT, *Psychology Department, New Mexico State University, Las Cruces, New Mexico 88003*

SYLVIA SHEPPARD, *Booz, Allen & Hamilton, Inc., Bethesda, Maryland 20814*

ELLIOT SOLOWAY, *Department of Computer Sciences, Yale University, New Haven, Connecticut 06520*

JOHN THOMAS, *IBM T. J. Watson Research Center, Yorktown Heights, New York 10598*

Preface

In reading the preface to a book, you could certainly get the impression that numerous people were involved in making it come into being. But it is probably not until you write or edit one, that you really appreciate the number of people that a published book actually requires.

There are a number of ways to describe the history of a book. In a very real sense the roots of every book are lost in antiquity with the proto-people who first communicated via language. In another way, a book is the product of a synergism resulting from the personal histories of all the people involved. In a less metaphysical sense, however, this book began one evening after an ANSI Meeting, and then took shape when a small group of people got together for dinner in Washington, D.C. after a conference on software metrics. We all shared the common belief that the major bottleneck to increased productivity in the use of computers was in the human factors—the usability.

It had been clear to us, individually, for some time that the drastic improvements in hardware price/performance meant that proportionally, the speed and effectiveness with which people could learn and then effectively use computers were of major importance. We seemed to feel that the time had come to make this fact more widely known. What should be done? Hold a conference on human factors and computer systems!

Hold a conference! What could be simpler? As is probably obvious, it is not simple. This book would not have been written without the many people who made the conference possible. After some searching, we found a home for the conference at the National Bureau of Standards in Washington, D.C. and a sponsor in the Washington chapter of the Association for Computing Machinery (ACM). As our plans for the conference grew, we decided that in addition to publishing proceedings of the conference, we would have a book of selected papers and special issues of the *Communications of the ACM* and of *Behavior, Instrumentation, and Technology.*

We anticipated having as many as one hundred attendees. In order to process the papers, we selected a program committee; it turned out that over one hundred and fifty papers were submitted. Nearly everyone, it seemed, had something to say about the human factors of computer systems. As the conference date of March 1982 drew near, it became obvious that we would have over five hundred attendees. In actuality, over nine hundred people attended the conference. Jean Nichols, general chairman of the conference, did a superb job in organizing and coordinating the various elements of the conference. The National Bureau of Standards and the Washington Chapter of the ACM did a remarkable job in accommodating such a large number, while minimizing problems. The people involved deserve special thanks, particularly Charles Bridges of ACM.

Support for the idea of a book came from John O'Hare of the Office of Naval Research Engineering Psychology Programs. Beyond that, however, John has supported much of the research in software human factors for many years, including much of our own earlier research. Special thanks goes to John for his foresight in realizing the importance of this area so early.

Thanks for hard work also goes to the members of the program committee and the reviewers of the papers in the conference, as well as the authors. On the program committee were: Michael Atwood, James Bair, Ruven Brook, H. E. Dunsmore, James Foley, Elihu Gerson, Thomas Green, Elizabeth Kruesi, David Lenorovitz, Thomas Moran, Franklin Moses, Peter Newsted, Phyllis Reisner, and Roger Schvaneveldt.

In order to publish a book, a publisher is also needed and we would like to thank Walter Johnson, founder and president of Ablex, for his continued support in this effort and the members of his staff.

We would also like to thank Ben Shneiderman, the editor of this series, not only for his help in this book, but even more important, for reminding numerous people at numerous conferences that computers are tools to serve people, and not vice versa.

Of course, the editors needed the support of our management to continue our efforts to ensure good human factors, not only in the products of our respective companies, but also in general. At Sperry, a number of people supported me [M. L. Schneider] in this endeavor. Dr. Richard Wexelblat, now at ITT, who encouraged my work in the conference, deserves special thanks. Dr. Hans C. Gyllstrom, director of Software Products Systems Design, has supported my work on this book. Mary Sterling, my secretary, is my right hand.

In my case [John Thomas], special thanks are due to Dr. Lewis

Branscomb, former head of the National Bureau of Standards and IBM's chief scientist, who supported my involvement with the conference and did much to bring the increased importance of human factors to the attention of all IBM. Special thanks are also due to Dr. Stephen Boies, manager of IBM's Office Applications Research and developer of IBM's Audio Distribution System. Our secretaries, Carol Johnson, Judy Cantor, and Ann Hubby, were also invaluable. Finally I would like to thank my wife Laura for her support in all things.

M. L. SCHNEIDER
JOHN THOMAS

Introduction

It is becoming obvious to more and more people that the tremendous advances in the computer industry in the past decades are heralding a revolution at least as profound as the industrial revolution. Hardware costs have plummeted and now, at least to many of us, the major challenge facing the computer industry is making computers that are easy to learn and easy to use. In other words, the question is, how can we make computer systems with good human factors?

While much is known about the physical aspects of human factors (e.g., Huchingson, 1981), relatively little is known about how to write software to maximize its usability. Certainly, given our current state of knowledge, it would be premature to claim that we even know "the" best way to study the problem of software human factors. In this book, therefore, a number of different approaches to various related problems are discussed.

Computers are difficult and often frustrating for new users. Yet video games, which seem to require at least as much in terms of information-processing capacity, are enjoyable and "natural." In Chapter 1, Thomas Malone empirically examines the factors that make video games enjoyable noting that some of these factors may be incorporated into more serious interfaces. "Heuristics for Designing Enjoyable User Interfaces: Lessons from Computer Games" discusses features of computer games that are important to enjoyment and how they might be incorporated into other user interfaces. Additional thinking along these lines can be found in Carroll and Thomas (1981) and Carroll (1982).

In Chapter 2, John Carroll and Robert Mack of IBM's T. J. Watson Research Laboratory study, in depth, the behavior of new users learning a real word processor. The investigators find ample evidence from their protocol analysis that learning cannot be validly viewed in this context as a passive process but is an active exploratory one. "Learning to Use a

Word Processor: By Doing, by Thinking, and by Knowing" includes an appendix with extended illustrative protocols of learners. Their work also illustrates that the study of humans learning and using computer systems is more than simply applying known psychological principles to the practical problem of making computer systems better. There is also a fundamental scientific motivation to this work.

Any behavioral scientist interested in human thinking and language would love to have the opportunity to observe over the course of a lifetime the evolution of human spoken speech or the evolution of written speech. The adaptation of human beings to these new media provides essential material about what is fundamental and what is adaptable in human thought and language. These opportunities have passed us by and we must reconstruct the history of the beginnings of spoken and written language.

We are, however, standing in the middle of a high-speed revolution of just as profound an impact—the computer evolution. We can observe first-hand how people learn to adapt to their purposes a new technology which allows exciting new possibilities for human beings to communicate thoughts across time and space via the computer. Watching this process intelligently cannot fail to give us a deeper understanding of what human thought really is, and, in a real sense, what it means to be human. In the study by Carroll and Mack, we are given insight into the nature of complex learning as well as what makes a system easy to learn.

Although it is certainly not trivial, in principle, we do understand how to measure how easy to learn and easy to use a system is once it exists. One measures the performance of real users doing real tasks on the real system. In the real world, however, by the time the real system exists, it is often too late for any fundamental changes (at least on that system).

Suppose that there were a way to predict which systems would be easy to learn and easy to use? And, suppose we could tie such a system of prediction into psychological and psycholinguistic theory? The work of Phyllis Reisner at IBM's San Jose Research Laboratories represents an attempt to accomplish this goal. In Chapter 3, Dr. Reisner describes a way of using formal grammar to predict what will be easy and hard about a system.

While new users are obviously a very important group to understand and design for, there is a large and growing group who do not simply use existing computer systems but who themselves design, implement, alter, test, and debug computer software. These "programmers" constitute a critically important group of users. In Chapter 4, investigators explore what makes programs and programming complex, what cognitive structures programmers use to simplify the task, and empirically investigate various documentation options. Daniel McNicholl and Ken

Magel present work on "Stochastic Modeling of Individual Resource Consumption During the Programming Phase of Software Department."

Kate Ehrlich and Elliot Soloway are interested in the knowledge structures that programmers use in programming. In Chapter 5, "An Empirical Investigation of the Tacit Plan Knowledge in Programming," one sees some rather clever ways of trying to get at these structures.

At least since Jerry Weinberg's (1971) seminal work, *The Psychology of Computer Programming*, it has been clear that it is important not only to write "efficient" code in terms of storage requirements and execution speed, but that it is also important to write understandable code. Often the programmer (or another programmer) must read and understand a program years after it was initially written. Sylvia Sheppard, John Bailey, and Elizabeth Kruesi Bailey present an "Empirical Evaluation of Software Documentation Formats" in Chapter 6.

Eventually, everyone may be a capable programmer. For now, however, a large number of users interact with computers at least partly via menus. Ricky Savage and James Habineck report in Chapter 7 on the use of empirical studies that improved the menu structures in the development of IBM's System/34. "A Multilevel Menu-Driven User Interface" illustrates that empirical work can cause changes in the development of real products.

Another approach to providing a computer interface for the naive user is that of natural language. An extensive program of research by George Furnas, Tom Landauer, Louis Gomez, and Susan Dumais at Bell Laboratories has explored the potential problems and partial solutions to having first-time users input their own words to describe desired actions and objects. "Statistical Semantics: Analysis of the Potential Performance of Keyword Information Systems" reports on this work.

In addition to menus and natural language, the third major approach to interfaces for non-DP users is in the graphical presentation of information. In Chapter 8, Timothy Goldsmith and Roger Schvaneveldt report on a series of experimental investigations of the effectiveness of various kinds of graphical displays in "Facilitating Multiple-Cue Judgments with Integral Information Displays."

The way that we communicate today in face-to-face communication, via phone, writing, or computer is largely determined by social habits built up over many thousands of years. In fact, our natural language capability was developed to allow communication under the constraints imposed primarily by our ability to produce sounds and secondarily by our abilities to write. With the advent of the printing press, film, television, and most especially, the computer, these limitations no longer dictate the structure of communication. Despite this potential freedom brought about by computers, however, most communication between human beings still proceeds in the slow linear fashion dictated by two stone-age humans shaping their mouths and vibrating their vocal folds.

We know from studies of sped speech, for example, that even without restructuring language, people can comprehend spoken text nearly twice as fast as people can spontaneously generate it. This suggests that with the advent of unlimited intelligible speech synthesis, spoken messages can be comprehended in half the time.

By the intelligent use of graphics and the proper selection of pictures from data banks of video images, entire arrays of data could be presented to people much more quickly than is possible by writing and speaking. We stand on the threshold of a new age brought about by new technology. The extent to which that new age represents progress in human productivity and enjoyment or merely a change of fashion without real progress depends heavily upon the work of investigators such as the ones who contributed to this volume.

References

Carroll, J. M. The adventure of getting to know a Computer. *Computer* (Nov. 1982), 49–58.
Carroll, J. M., & Thomas, J. C. Metaphor and the cognitive representation of computer systems, IEEE Transactions on Systems Man and Cybernetics, SMC-12(2), 107–116, 1982.
Huchingson, R. D. *New horizons for human factors in design.* New York: McGraw-Hill, 1981.
Weinberg, G. M. *The psychology of computer programming.* New York: Van Nostrand, 1971.

1

Heuristics for Designing Enjoyable User Interfaces: Lessons from Computer Games*

THOMAS W. MALONE

This chapter discusses two questions: (a) Why are computer games so captivating? and (b) How can the features that make computer games captivating be used to make other user interfaces interesting and enjoyable? First, several studies of what makes computer games fun are summarized. Then a set of guidelines for designing enjoyable user interfaces is developed. The guidelines are organized in three categories: *challenge, fantasy,* and *curiosity*.

Programs are *challenging* when they suggest clear, personally meaningful goals whose outcomes are uncertain because of variable difficulty levels or multiple levels of goals. Programs can engage *fantasy* by including emotionally appealing fantasies, preferably ones that provide useful metaphors with physical or other systems the user already understands. Finally, programs can stimulate *sensory curiosity* by using audio and visual effects, and *cognitive curiosity* by introducing new information when users feel their current knowledge is incomplete, inconsistent, or unparsimonious. A number of examples of applying these principles to designing computer programs are included.

In this chapter, I will discuss two questions: (*a*) Why are computer games so captivating? and (*b*) How can the features that make computer games captivating be used to make other user interfaces interesting and enjoyable to use?

After briefly summarizing several studies of what makes computer games fun, I will discuss some guidelines for designing enjoyable user interfaces. Even though I will focus primarily on what makes systems enjoyable, I will suggest how some of the same features that make systems enjoyable can also make them easier to learn and to use.

*This paper appeared in the *Proceedings of the Conference on Human Factors in Computer Systems*. Gaithersburg, MD, March 15–17, 1982, published by the Association for Computing Machinery.

STUDIES OF ENJOYABLE COMPUTER GAMES

To help determine what makes computer games so captivating, I conducted three empirical studies of what people like about the games. All of these studies are described in more detail elsewhere (Malone, 1980, 1981a, 1981b) and are only briefly summarized here. The primary purpose of these studies was to help design highly motivating instructional environments, but they also have important implications for designing other user interfaces.

Darts

To illustrate the methodology used, I will briefly describe one of the studies. This experiment analyzed a game called Darts that was designed to teach elementary students about fractions (Dugdale & Kibbey, 1975). In the version of the game used, three balloons appear at random places on a number line on the screen and players try to guess the positions of the balloons (see Figure 1). They guess by typing in mixed numbers (whole numbers and/or fractions), and after each guess an arrow shoots across the screen to the position specified. If the guess is right, the arrow pops the balloon. If wrong, the arrow remains on the screen and the player gets to keep shooting until all the balloons are popped. Circus music is played at the beginning of the game and if all three balloons in a

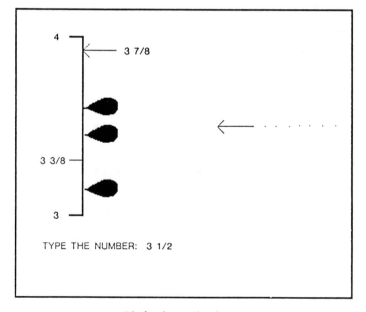

FIGURE 1. **Display format for the Darts game.**

round are popped in four tries or fewer, a short song is played after the round.

To find out what features contribute most to the appeal of this game, I constructed eight different versions of the game by taking out, one at a time, features that were presumably motivational. The features removed included: the music, the scorekeeping, the fantasy of arrows popping balloons, and different kinds of feedback (see Figure 2).

Eighty fifth-grade students were each assigned to one of the eight versions and then allowed to play with either their version of Darts or with a version of Hangman that was the same for all students. The primary measure of appeal of the different versions was how long students played with their version of Darts in comparison to Hangman. This measure was also highly correlated with how well students said they liked the game at the end ($r = .30, p < .01$).

Somewhat surprisingly, there was a significant difference between boys and girls in what they liked about the game. An analysis of variance of the time spent playing Darts revealed significant effects of condition ($F(7,48)=2.21, p < .05$) and of the sex by condition interaction ($F(7,48) = 4.84, p < .001$). A detailed interpretation of the differences (shown in Table 1) is given in (Malone, 1980), and (Malone, 1981b). Briefly, the girls' dislike of the intrinsic fantasy of arrows and balloons (Condition 7 vs. 8) appears to be because they dislike the arrows and balloons fantasy in the first place and the fantasy is more salient in the intrinsic than the extrinsic version. Furthermore, the differences between Conditions 3 and 4 for boys and between Conditions 5 and 6 for girls appear to be less reliable than the others, because they are not significant when the time measures are scaled according to a plausible model of choice behavior.

In summary, the primary result of this experiment was that boys like the fantasy of arrows popping balloons, and girls appeared to dislike this fantasy. The results also showed that fantasy made more difference in the appeal of the game than did simple feedback. In other words, even though responsiveness is often mentioned as an important reason why computers are captivating, the simple feedback in the game was not as important as the fantasy in making this game fun.

I think the most important implication of this experiment is that fantasies can be very important in creating intrinsically motivating environments but that, unless the fantasies are carefully chosen to appeal to the target audience, they may actually make the environment less interesting rather than more.

Other Studies

Another game, called Breakout, was studied in a similar way. In this game, the player controls a paddle and tries to hit a ball so that it knocks

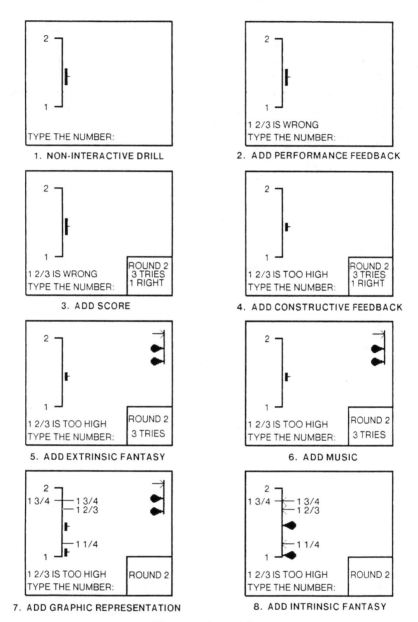

FIGURE 2. **Different versions of the Darts game.**

TABLE 1
Interest in Different Versions of the Darts Game

Condition	Time Playing Darts (0–40 min)	
	Boys	Girls
1. Noninteractive drill	20.5	15.5
2. Add performance feedback	18.8	20.2
3. Add scoring	24.2	19.8
4. Add constructive feedback	16.2*	22.2
5. Add extrinsic fantasy	25.8*	20.8
6. Add music	21.8	30.0*
7. Add graphic representation	28.3	29.8
8. Add intrinsic fantasy	34.5	19.8**
Average	23.4	22.0

$*p < .05$, for comparison with previous condition.
$**p < .01$, for comparison with previous condition.

all the bricks out of a wall. The visually compelling goal of knocking bricks out of the wall was found to be the most important of the features varied in this game. Finally, in a survey of the computer game preferences of 65 elementary school students, the features that were most strongly correlated with game popularity were the presence of an explicit goal, score-keeping, audio effects, and randomness.

IMPLICATIONS FOR DESIGNING ENJOYABLE USER INTERFACES

In this section, I will outline a general framework for analyzing the appeal of computer systems based on three categories: challenge, fantasy, and curiosity (see Table 2). The primary purpose of this framework is to serve as a checklist of heuristics for designing enjoyable user interfaces. One purpose of this paper is to show how this framework, which was developed elsewhere (Malone, 1981b) for analyzing instructional environments, can be applied to more general user interfaces.

The motivational processes discussed in this chapter are, in many ways, less well-understood and subject to much larger individual differences than many of the cognitive processes involved in human–computer interactions. Accordingly, the heuristics in this section should be viewed as suggestions, not as requirements. Many of them are only appropriate for some people in some situations and they must be applied with care.

TABLE 2
Heuristics for Designing Enjoyable User Interfaces

  I. Challenge
    A. *Goal.* Is there a clear *goal* in the activity? Does the interface provide *performance feedback* about how close the user is to achieving the goal?
    B. *Uncertain outcome.* Is the outcome of reaching the goal uncertain?
      1. Does the activity have a *variable difficulty level?* For example, does the interface have *successive layers of complexity?*
      2. Does the activity have *multiple level goals?* For example, does the interface include *scorekeeping?*
  II. Fantasy
    A. Does the interface embody *emotionally appealing* fantasies?
    B. Does the interface embody *metaphors* with physical or other systems that the user already understands?
  III. Curiosity
    A. Does the activity provide an *optimal level of informational complexity?*
      1. Does the interface use *audio and visual effects:* (a) as decoration, (b) to enhance fantasy, and (c) as a representation system?
      2. Does the interface use *randomness* in a way that adds variety without making tools unreliable?
      3. Does the interface use *humor* appropriately?
    B. Does the interface capitalize on the users' desire to have *"well-formed"* knowledge structures? Does it introduce new information when users see that their existing knowledge is: (1) *incomplete,* (2) *inconsistent,* or (2) *unparsimonious?*

Toys and Tools

It is important in describing this framework, to distinguish two different uses of computing systems: *toys*—systems used for their own sake with no external goal (e.g., games), and *tools*—systems used as a means to achieve an external goal (e.g., text editors, programming languages, etc.).

As discussed below, good toys and good tools are similar in the ways they can use fantasy and curiosity, but in an important way they are opposite with respect to their requirements for challenge. Since most user interfaces are for tools, not toys, much of the motivation for using the system depends on the user's motivation to achieve the external goal. In cases where the external goal is not highly motivating (e.g., is routine and boring), the toylike features discussed below can be especially useful in making the activity enjoyable.

Challenge

Goal

For an activity to be challenging, it needs to have a *goal whose outcome is uncertain.* As described above, computer games without explicit or easily generated goals were less enjoyable than games with goals. In other words, a challenging toy must either build in a goal or be such that users can easily create their own goals for its use. A good tool, on the other hand, is designed to achieve goals that are already present in the external task.

For both toys and tools, however, users need some kind of *performance feedback* to know how well they are achieving their goals. In games, this performance feedback is provided by things like the missing bricks in Breakout and the position of the incorrect arrows on the number line in Darts. There may be similar ways to incorporate performance feedback for the external task into tools. For example, the Writer's Workbench developed at Bell Laboratories (MacDonald, Frase, Gingrich, & Keenan, 1982) measures various stylistic features of manuscripts, such as word length, sentence length, percentage of sentences using passive voice, and so forth. These rudimentary kinds of performance feedback for the external goal of producing a readable manuscript may enhance the challenge of using the tool.

Uncertain Outcome

The most important difference between toys and tools occurs with respect to the uncertainty of outcome of reaching a goal. If a user is either certain to achieve a goal or certain *not* to achieve it, the activity will not be very challenging. For an activity to be challenging, the outcome of achieving the goal must be uncertain.

One way of making the outcome of a computer game uncertain for a wide range of players, or for the same player over time, is to have a *variable difficulty level*. For example, in the Breakout game, after a player hits the ball correctly five times in a row, the ball speeds up. As Nolan Bushnell, the founder of Atari, Inc., has been quoted as saying, "A good game should to easy to learn, but difficult to master."

A good tool, on the other hand, should be both easy to learn and easy to master. Since the outcome of the external goal (writing a good letter, getting a program to work) is already uncertain, the tool itself should be reliable, efficient, and usually "invisible." In other words, the tool users should be able to focus most of their attention on the uncertain external goal, not on the use of the tool itself. In a sense, a good game is intentionally made difficult to play, but a tool should be made as easy as possible to use. This distinction helps explain why some users of complex systems may enjoy mastering tools that are extremely difficult to use. To the extent that these users are treating the systems as toys rather than tools, the difficulty increases the challenge and therefore the pleasure of using the systems.

In spite of the differences between toys and tools, there is a way tools can use variable difficulty levels to increase challenge and, at the same time, probably improve learnability as well. I have heard many system design arguments in which the fundamental conflict is between, on the one hand, a desire to have the system be simple and easy to learn for beginning users, and on the other hand, the desire to have it be powerful and flexible for experienced users. Many of these arguments could be resolved by consciously building in a logical progression of *increasingly complex microworlds* for users at different levels of expertise (Fischer, Burton, & Brown, 1978).

For example, a multilayered text editor could be designed so that beginning users need only a few simple commands and more advanced users can use more complicated and more powerful features of the system. Ideally, this system should be internally consistent at each level so that the error messages for users of the first level would never assume any knowledge of concepts used only in more advanced levels. In fact, some commands that might make sense if made by an advanced user should probably be treated as errors if made by a beginning user.

The point here is that a multilayered system could not only help resolve the trade-off between simplicity and power, it could also enhance the challenge of using the system. Users could derive self-esteem and pleasure from successively mastering more and more advanced layers of the system, and this kind of pleasure might be more frequent if the layers are made an explicit part of the system.

Another way of providing uncertain outcome in computer games is to have *multiple level goals* all present in the environment at the same time. For example, in the Breakout game, long before there is any hope of a beginning player breaking out all the bricks, the player can still be challenged by lower level goals like breaking out any brick in the third row or breaking out all the bricks in the first row. Or in the Darts game, players who are certain they can pop all the balloons can still be challenged by trying to pop all the balloons in as few tries as possible. In general, *score-keeping* and *timed responses* are two common ways of enhancing multiple level goals in computer games.

It may be possible to incorporate similar kinds of multiple level goals into tools as well. For example, I think some users of a text-editing system would be challenged by having the system automatically maintain scores like typing speed or number of corrections made. If the text-editing task is boring or routine for the user, this challenge might increase the pleasure of using the system. (It would almost certainly *not* increase the pleasure of using the system, however, if such scores were used for surveillance by organizational superiors.)

Another way of providing multiple level goals in a system is by having a lot of user programming capabilities. If users can write procedures to do subcomponents of their routine on-line tasks, then they can continue to be challenged by trying to make their system more efficient for the tasks they do. For example, if a text-editing system allows users to define their own macros, people who prepare many similar documents can be challenged by constructing macros to make this process more efficient.

Fantasy

Fantasy is probably the most important feature of computer games that can be usefully included in other user interfaces. By a system with fantasy, I mean a system that evokes mental images of physical objects or social situations that are not actually present. For example, the Breakout and Darts games evoke images of physical objects like balls, bricks, darts, and balloons; and the omnipresent computer Adventure game evokes images of caves, dwarfs, birds, and so forth. I think fantasies have two important aspects for designing user interfaces: *emotions* and *metaphors*.

Emotions

Fantasies in computer games almost certainly derive some of their appeal from the emotional needs they help to satisfy in the people who play them. It is very difficult to know what emotional needs different people have and how these needs might be partially met by computer games. As the Darts experiment described above suggests, there are large differences among people in what fantasies they find appealing. Designers of computer systems that embody fantasies should either be very careful to pick fantasies that appeal to their target audience or they should provide several fantasies for the same system so that different people can select different fantasies.

One use of fantasy in computer systems might be to give different "personalities" to different parts of a system. For example, the operating system might have one personality, different application programs might have other personalities, and file servers on a network might have still other characteristics. But the personalities of the different parts of the system could be different for different users. Some users might like to work in a world of wizards, dragons, and trolls, others might prefer a world of dogs, cats, and rabbits, or even a world populated by characters from Star Trek or Charlie's Angels. Not only could these fantasies increase the emotional appeal of the systems, they could also be useful metaphors to help users learn the difference between different parts of a system—something that is not at all trivial for beginning users.

Carroll and Thomas (1980) suggest another use of fantasy in "reframing" routine information processing tasks to make them more interesting. For example, they suggest that certain kinds of factory control operations (e.g., monitoring a steam engine) could be presented to the user as more captivating "virtual tasks," such as flying an airplane full of passengers onto a dangerous landing field. Measurements in the factory control space could be translated into the airplane metaphor, and actions taken in the airplane fantasy could be translated into actions in the factory.

This kind of reframing would presumably be appropriate only if the original task was boringly routine. Fantasies could make such routine tasks more enjoyable. But unless the outcome of reaching the goal is made uncertain (e.g., with an adjustable difficulty level or multiple level goals), the fantasy tasks could become boring as well.

Metaphors

In addition to being emotionally appealing, fantasies that are analogous to things with which the users are already familiar, can help make the systems easier to learn and use (see Carroll & Thomas, 1980 and Halasz & Moran, 1982 for extended discussions of this point). For example, I think one of the reasons for the popularity of the VisiCalc system (Bricklin & Frankston, 1979) is the fact that the program is very analogous to the kind of paper "spread sheet" that was already widely used by many of the business analysts who purchased VisiCalc systems.

The user interface for the Xerox Star workstation is another example of a system that makes extensive use of metaphors. Much of the manipulation of information takes place by moving icons around on a "desktop" that is simulated on the screen. The icons are pictorial representations of familiar objects like in-baskets, file folders, and filing cabinets. To the extent that this fantasy is analogous to real desktops, it presumably makes the system easier to learn and use.

Curiosity

The final category of features that make computer games appealing includes features that evoke the users' curiosity. Environments can evoke curiosity by providing an *optimal level of informational complexity* (Berlyne, 1965, and Piaget, 1951). In other words, the environments should be neither too complicated nor too simple with respect to the user's existing knowledge. They should be *novel* and *surprising*, but not completely incomprehensible. In general, an optimally complex environment will be one where the learner knows enough to have expectations about what will happen, but where these expectations are sometimes unmet.

One important way computer games evoke what might be called

sensory curiosity is by using audio and visual effects. Audio and visual effects can be used (*a*) as decoration, (*b*) to enhance fantasy, and, perhaps most importantly, (*c*) as a representation system. Examples of using audio and visual effects as representation systems include (*a*) using different tones for errors and for successful entries (*b*) using graphs instead of numbers, and (*c*) using icons to represent different parts of a system (such as in-baskets and out-baskets) and different commands.

Randomness and *humor,* if used carefully, can also help make an environment optimally complex. As a simple example, one computer system at Stanford ends each terminal session with a randomly chosen saying, often resembling a fortune from a Chinese fortune cookie. Such features seem likely to increase the enjoyment of using a system, but great care must also be exercised—especially when introducing humor—to avoid inappropriate (or unhumorous) additions. For example, if randomness is used in a way that makes tools unreliable it will almost certainly be frustrating rather than enjoyable.

Curiosity can also be thought of as a drive to bring "good form" to knowledge structures. In particular, people try to make their knowledge structures *complete, consistent,* and *parsimonious,* and one can evoke curiosity by making people think their current knowledge is incomplete, inconsistent or unparisimonious. Computer system designers can take advantage of this principle by, for example, introducing new features of a system only when users see a need to do something they don't know how to do (i.e., see an incompleteness in their knowledge) or where they can do something with fewer steps (e.g., more parsimoniously) than they have previously done it.

CONCLUSION

Table 2 lists the major features of computer games I have discussed that can be incorporated into other user interfaces. This table should be viewed as a checklist of ideas to be considered in designing new interfaces. Certainly not all the features will be useful in all interfaces. But I think that many user interfaces could be improved by systematically considering the inclusion of features such as multiple layers of complexity, productive and involving metaphors, and useful sound and graphics.

It is, of course, easy to use these features badly. It would be very easy, for example, to build user interfaces that include garish graphics, inappropriate fantasies, and sick humor. But with creativity and strong aesthetic and psychological sensitivity, I think the pervasive computer systems of tomorrow can be made not only easier and more productive to use, but also more interesting, more enjoyable, and more satisfying.

ACKNOWLEDGMENTS

I would like to thank Tom Moran, Kurt VanLehn, Bert Sutherland, and Austin Henderson for helpful comments that are included in this discussion.

REFERENCES

Berlyne, D. E., *Structure and direction in thinking.* New York: Wiley, 1965.

Bricklin, D., & Frankston, B. *VisiCalc computer software program.* Sunnyvale, CA: Personal Software, 1979.

Carroll, J. M., & Thomas, J. C. *Metaphor and the cognitive representation of computing systems.* Yorktown Heights, NY: IBM Watson Research Center technical report no. RC 8302, 1980.

Dugdale, S. & Kibbey, D. *Fractions curriculum of the PLATO elementary school mathematics project.* Computer-based Education Research Laboratory Technical report. University of Illinois, Urbana, IL, 1975.

Fischer, G., Burton, R. R., & Brown, J. S. Aspects of a theory of simplification, debugging, and coaching. Proceedings of the *Second Annual Conference of the Canadian Society for Computational Studies of Intelligence.* Also available as Bolt Beranek and Newman, Inc. Technical report no. 3912 (ICAI Report No. 10), Cambridge, MA, 1978.

Halasz, F., & Moran, T. P. Analogy considered harmful. Proceedings of the conference on *Human Factors in Computer Systems.* Gaithersburg, MD, March 15–17, 1982.

Macdonald, N. H., Frase, L. T., Gingrich, P. S., Kennan, S. A. The writer's workbench: Computer aids for text analysis. *IEEE Transactions on Communications COM-70,* 105–110, 1982.

Malone, T. W. What makes things fun to learn? A study of intrinsically motivating computer games. Ph.D. dissertation, Department of Psychology, Stanford University. Also available as technical report no. CIS-7 (SSL-80-11), Xerox Palo Alto Research Center, Palo Alto, CA, 1980.

Malone, T. W. What makes computer games fun? *Byte, 6,* 258–277, 1981 (December). (a)

Malone, T. W. Toward a theory of intrinsically motivating instruction. *Cognitive Science 4* 333–370, 1981. (b)

Piaget, J. *Play, dreams, and imitation in childhood.* New York: Norton, 1951.

2

Learning to Use a Word Processor: By Doing, by Thinking, and by Knowing

JOHN M. CARROLL
ROBERT L. MACK

Learning to use a word processor provides a study of real complex human learning that is fundamentally "active," driven by the initiatives of the learner—which are, in turn, based on extensive domain-specific knowledge and skill, and on reasoning processes which are systematic yet highly creative. State-of-the-art application systems and their training materials presuppose "passive" learners and are—to that extent—unusable. Some general design implications of active learning were discussed.

Suppose that people learned "passively." That is, suppose that designers of systems and of their training materials could place new users in a carefully programmed learning environment with confidence that through this experience new users would become skilled experts. These passive learners would presumably read descriptions and explanations of the system's functional capabilities; they would carefully follow exercises to drill and develop their skill and understanding; and gradually these experiences would converge on a mature stereotype of skill and understanding.

In point of fact, the above is pretty much what designers of state-of-the-art application systems and training material have supposed. Were this supposition correct, the human factors of computer systems would be in good shape today: the technology for "passive" learning of systems is well developed. But unfortunately, the supposition is almost totally wrong. As we will argue here, people learn *actively* not passively. The problems new users have with existent systems and training are testimony to this contention. That people do learn to use such systems via such training programs is testimony to the adaptability and intellectual tenacity of people. This point will also be amply illustrated in our discussion.

In order to be more usable, future application systems and their training support will need to accommodate the real (= "active") learner, rather than what might have been—from the perspective of the system designer and manual writer—the ideal (= "passive") learner. Toward the end of this chapter we will sketch some general design proposals along these lines and attempt to theoretically bring our work to bear on the analysis of human behavior and experience in realistic situations that is developing in current Cognitive Science (Norman, 1981).

METHOD AND OVERVIEW

The research discussed consists of studies of office personnel learning to use word processing equipment. In this chapter, we are not concerned with issues regarding particular systems (see Mack, Lewis, & Carroll, 1983). Our principal focus will be on the learning strategies (actually the classes of learning strategies) we have identified in the course of this work. The three classes of strategies we address are learning by doing, learning by thinking, and learning by knowing.

Our view is that these are natural—albeit complex—strategies for people to adopt when confronted by a learning task of nontrivial complexity. The learners we have studied are almost entirely "innocent" with respect to computer technology. In the context of learner innocence, we argue, these "natural" strategies entail severe and wide ranging learning problems. Analysis of these problems, in turn, suggests research directions for the analysis of real human learning within contemporary cognitive science and practical directions in which computer systems, and the educational technologies that support their training and use might evolve.

The Learners and Their Task

In this research project, ten office temporaries spent four half-days learning to use one of two possible word processing systems in our laboratory. Four were given a subset of the word processing operations for a general purpose, command-based system. The other six were asked to learn basic functions of a commercially available word processor which was menu-based.

These people were highly experienced in routine office work, but quite naive with respect to computers in general and word processing systems in particular. We asked them to imagine a scenario in which a word processing system had recently been introduced to their office, and they had been asked to be the first to learn it (to then pass this knowledge on to colleagues). The point was that they were to learn to use the system using the training materials that accompany it as their only resource. The manual we used was designed to be used as the nucleus of a "self-study" training method, although, in fact, such material is rarely used in complete isolation from other methods such as tutorial assistance (which is not surprising in light of our results, see below).

The materials we asked our learners to study addressed basic letter entry and revision skills (including formatting, printing, document retrieval, and document merging), and, in the case of the commercially available system, the use of command menus, the interpretation of messages, and the use of the on-line Help facility.

Learners spent varying amounts of time with this task, but we stopped them in any case after about 12 hours in order to ask them to try a simple "transfer task," involving typing in, revising, formatting, and then printing out a one-page letter. This transfer task was not part of the self-study manual; it served as a sort of benchmark achievement test. Finally, we asked learners about their prior work background, about specific queries that occurred to us during the course of the learning sessions, and about their overall impressions of the system and the experience.

Our concern here is not to detail the results of the project; they can be found in Mack et al. (1983). Suffice it to say that a great many problems cropped up during the sessions and that even after the 12 hours or so of self-study learning not a single learner was able to complete the transfer task without some kind of serious difficulty in editing subtasks such as document creation, retrieval or printing, or in actually entering and revising text.

Two very general points should be stressed about this project. First, the domain in which we studied learning is drawn from the real world of actual learning that real people must cope with: we studied real word processing systems, real training books, and users drawn from the population to whom the system and the books were targeted. We take this project to constitute an "ecologically valid" investigation of human learning. Second, the systems and the manuals which were our materials for this work are state-of-the-art. Thus, our findings—especially those pertaining to learner problems—are not abstract effects of "materials," but rather have immediate and specific practical implications for the design of technology in the word processing domain (see Mack et al., 1983).

Thinking Aloud

Our method involved prompting learners to "think aloud" as they worked through the training materials. They were to report questions that were raised in their minds, plans and strategies they felt they might be considering or following out, and inferences and knowledge that might have been brought to awareness by ongoing experiences. We remained with the learners to keep them talking and to intervene, if at any time it appeared that a problem was so grave that a learner might leave the experiment if we did not help out. Our prompting remained nondirective, and indeed once learners got going we needed to prompt very infrequently at all. We had to actually intervene rarely.

Throughout the learning sessions, video and sound recordings were made (in what we hoped was an unobtrusive manner). Our analysis consisted first of an enumeration of "critical incidents" which were cataloged and classified in various ways. This was constrained by the consensus of the three experimenters (the authors and Clayton Lewis). The chief goal of this was to form a picture of the typical experience of a learner, and it is this induced "prototype" learning experience to which we will refer in what follows (for details, see Mack et al., 1983). Our method of reporting will be to cite induced generalizations and to illustrate these by concrete examples transcribed from our audio recordings.

We are aware that the thinking aloud technique has been of great methodological interest recently, but this topic lies outside the scope of the present study (see Ericsson & Simon, 1980; Nisbett & Wilson, 1977).

How People Learn

Perhaps the most apt description of the world of the new user of a word processor is that often quoted phrase of William James: "a bloomin' buzzin' confusion." People in this situation see many things going on, but they do not know which of these is relevant to their current concerns. Indeed, they do not know if their current concerns are the appropriate concerns for them to have. The learner reads something in the manual; sees something on the display; and must try to connect the two, to integrate, to interpret. It would be unsurprising to find that people in such a situation suffer conceptual—or even physical—paralysis. They have so little basis on which to act.

And yet people do act. Indeed, perhaps the most pervasive tendency we have observed is that people simply strike out into the unknown. If the rich and diverse sources of available information cannot be interpreted, then some of these will be ignored. If something *can* be interpreted (no matter how specious the basis for this interpretation), then it will be interpreted. Ad hoc theories are hastily assembled out these odds and ends of partially relevant and partially extraneous generalization. And these "theories" are used for further prediction. Whatever initial confusions get into such a process, it is easy to see that they are at the mercy of an at least partially negative feedback loop: things quite often get worse before they get better.

Designers of word processors and training technology probably would have liked things to have been different. The easiest way to teach someone something is, after all, to tell them. However, what we see in the learning-to-use-a-word-processor situation is that people are so busy trying things out, thinking things through, and trying to relate what they already know (or believe they know) to what is going on that they often do not notice the small voice of structured instruction crying out to them from behind the manual and the system interface.

What's wrong? We would argue that the learning practices people adopt here are typical, and in many situations adaptive. The problem in this particular learning situation is that new learners of word processors are *innocent* in the extreme. Each feature of a word processing system may indeed have a sensible design rationale from the viewpoint of the systems' engineer, but this rationale is frequently far beyond the grasp of the new user. "Word processor," so far as we know, is not a natural concept. People who do not know about word processors have little, possibly nothing, to refer to in trying to actively learn to use such things. Innocence turns reasonable learning strategies into learning problems. This is what this study is about.

LEARNING BY DOING

Our learners relentlessly wanted to learn by trying things out rather than by reading about how to do them. In part this was impatience: they were reluctant to read a lot of explanation or get bogged down following meticulous directions. But it also devolved from mismatched goals: Learners wanted to discover how to do specific things at particular times, and this did not always accord with the sequence in which topics were treated in the manual.

Jumping the Gun

Half of our learners impatiently tried to sign on to the system before reading how to do so. Table 1 presents highlights of one such learner. The learner had reached a point in the manual which showed the sign on display. This was intended as orientation and included no instruction about how to sign on.

The learner immediately incurs a number of errors which prevent her from signing on easily, and she is forced to try out various actions on her own initiative in order to deal with these ancillary problems. Indeed, it is only after considerable exploration of keys and commands that she is able to sign on. The specific character of her errors, and the exploration she engages in reveal a profound lack of knowledge about computers.

TABLE 1
Signing on the Word Processing System

(1) *Learner has been trying to type the letter string "abc" on display in order to observe cursor movement and practice backspacing. She is typing this string on the operator name field for the sign-on menu. Due to a number of errors which include prematurely trying to sign-on with "abc" as the incorrect—and inadvertent—operator name, the learner decides to simply start over.*

> L: So now let's see. Could I turn it off and start all over? That's what I would do. Will it hurt anything?
> E: You're in control.
> L: Let's see what happens. If I were all alone that's what I would do.
> E: Yup! You're all alone!

> *Powers off, and immediately back on. A new sign-on menu display appears.*

(2) *Correctly types operator name "learner6" in the sign-on menu field labeled "operator name."*

> L: Uh, the six [*i.e., of "learner6"*] was right next to it [*i.e., no space*] and it entered without typing anything. Do I have to make it, press "enter" to make it stay? No. Didn't say anything about that so we will forget about that.

> *Reference unclear, but she does not "enter" the filled-in sign-on menu. Instead, she continues with the next exercise, involving typing the letters "abc" and backspacing. This is intended to be done on the sign-on menu to illustrate how one can correct mistakes made while filling in menu parameters.*

(3) L: Okay now we are off. [*Reads aloud*] "If you have worked with this type "abc" and watch the cursor move."

> *Now she types "abc" right after the operator name which causes an error signalled several ways: audio message, a message about exceeding the length of the menu field, an indicator light signaling the need to take corrective action (labeled "reset/help"), and a locked keyboard which can only be unlocked by pressing a reset key.*

> L: Whoops!
> E: What do you think happened then?
> L: A bell rang. A buzzer or something.

(4) L: *Reads instructions on Sign-on menu display.* "Enter to finish" Okay let's see what happens. [*Presses the enter key*]

> E: You pressed enter?
> L: I pressed enter and nothing happened. [*The keyboard is locked*]

> *Reads more fragments of manual, stops at: "Ask your training administrator for help."* How do I get the menu? Request the menu again. All right. Press request, cancel.

> *Presses keys labeled request and cancel, to no effect because of the locked keyboard. Reference to "the menu" is not clear but may reflect her current understanding that she has not signed on yet or that she needs some other kind of display to proceed.*

> Okay. I didn't get the original menu to press anything.

> *Learner believes she needs a different menu, one which lists the tasks or commands available (e.g., create or retrieve or print a document). It appears after one completes sign-on.*

> Oh "enter" it. Nothing.

TABLE 1—*Continued*

Presses enter, which would have the effect of bringing up another display, but to no effect because of locked keyboard.

Now presses request key which is not relevant to anything here.

E: You pressed request?
L: Request and enter, trying to find that menu page because this page is not; I don't know how to get off this page. *Not clear what she is referring to here, but clearly she believes that she must have a different display to work on.*

All right. Come on. *Reads more fragments of manual referring to functions which advance and return the function from one menu parameter field to another.*

Okay. Pressed that.
E: Variable advance?
L: Right. "Watch the cursor . . ." Nothing. [*Keyboard is still locked.*] Great!
E: Okay what are you pressing there?
L: Variable Advance. [*then*] "Enter" to see if anything different might happen. Okay. Let's try it once more. Nothing!

(5) Still nothing. Okay. If I wanted to go back I would use the . . . ". . . look on the front . . .". Where is the "code" key? Now I'm trying the variable advance key and the code key to see if anything happens. Nothing happens! For all the punching I'm doing nothing at all is happening.

E: Do you have any idea why that might be?
L: Ah, I can't understand why it isn't entering, even entering the mistakes if I'm making mistakes. It's coming up with nothing. They did not tell me to get off this page.

New Audio Tape—some interaction may have been lost.

Now I'm touching the backspace key and it is not erasing anything. [*because keyboard remains locked*] So I have made some kind of error that has locked everything in here.

E: Nothing seems to be working?
(6) L: Right. Reset/Help? [*Presses the reset key, which is the correct remedy*]
E: Pressing reset?
L: I pressed reset/help to see what would happen there.
E: What do you think happened?
L: Nothing, really. [*In fact, the reset/help indicator light went off, signaling the unlocking of keyboard.*]
E: Why did you press that?
L: I'm trying to get back to a beginning. To start over.

All right. What would happen if I put it on "test"? *Referring to a button on the display unit which is labelled "test status." It is used to diagnose problems with the unit. It is totally irrelevant here.*

Turns test status "on" and a diagnostic pattern overlays screen.

Beautiful!

E: What happened?

(continued)

TABLE 1—*Continued*

L: I decided to put it on test. Status is "test" rather than normal. Now let's see what happens if you backspace.

Backspace works and she does so, deleting the incorrect string "learner6abc."

Oh, my gosh, are you sure this isn't hurting everything. Oh, hey, hurrah! Good! Very good! Now we are in business.

Now if we backspace, it takes it off. Okay. Fine. Now where did it tell me that I should have started at "test"?

(7) [*Sometime later the learner is reading the manual.*]
E: Can you summarize what you were just doing there?

L: Well I was looking at this [*page*] 1–8. I was looking at this diagram and it says "The reset/help indicator is on." So then I figured out that these are indicators. I didn't know that before this. *Summarizes what other indicator lights mean.*

Now I pressed the correct key and nothing happened, which is exactly what they tell you in the directions will happen. "When you are in a reset [*condition*] you cannot move the cursor or type anything." Now this was my probem before. But I wasn't up to page 1–9 so I didn't know that if the reset/help key, uh indicator is on the machine is going to do nothing. So you need to press . . .

E: Was that your problem before?
L: Yes.
E: Huh?
L: I didn't realize that it would not act as long as that was on, that I have to do something to get it back to acting again.
Okay. To be able to continue you need to press the reset key. So I press that. Now the . . . did they call it a code? An indicator. The indicator went off. The reset/help indicator went off. So evidentally we are back in business.

(8) Well, um. Do you want . . . maybe I should go back and put it on normal and see if I could do the same thing?

E: Well, you're in control.

Puts test status button on "normal" setting. Now it's on "normal." Let's do everything we did and see if it will work on this because we may still have been on the reset/help button over there. I don't know.

We'll type the "abc" [*types "abc" in the operator name sign-on menu field*] and see what happens. Then we'll try to backspace [*does so, deleting the "abc" she has just typed*] Does that alright. We'll go backwards and do the things they told us to do.

Tries function to advance the cursor to a new menu field. Which moves the cursor. Now they want us to move back. We'll code it in and we're back. [*i.e., when user uses cursor advance function in conjunction with a code key, the cursor returns or moves up to the next menu field*]

Okay. I'm running back through the instructions and it is working with it set on "normal." [*Reads fragments of manual—repeats exercise on making "invalid key" errors and pressing reset*] Okay I could have done the exercise on "normal."

E: Is that what you conclude?
L: That's what I concluded. Evidentally when I heard the bell before and I didn't

TABLE 1—*Continued*

know that the reset/help indicator was on or didn't realize what was happening in that column. In other words I got into trouble before I got to [*page*] 1–8! So now we know why nothing was registering because I hadn't read on page 1–9 that I had to press the reset key.

Okay. Now I am going on to exercise two. We're going to find out how to sign on.

At this point begins following directions to sign-on.

For example, in segment 2 the learner has typed her operator identification but she does not immediately realize that she must press the "enter" key to execute the sign on task. When she later tries the enter key (segment 4) she has in the meantime incurred an additional error that has locked the keyboard (segment 3). Hence, enter has no effect after all. In segment 5 she begins to appreciate that there is a problem after trying out a number of other function keys to no avail (e.g., a backspace to delete the erroneous typed parameter entry). Finally, in segment 6 the learner manages to solve the locked keyboard problem, but she confuses that action with an immediately following—and perhaps some salient—operation (involving a "Test Status" key) and so misinterprets the solution to her problem.

Jumping the gun is not always traceable to "mere" impatience. Often learners strike out on their own because their specific goals do not match those implied by the manual designers. As the first few segments of Table 2 reveal, the learner is not certain of the point of the exercise. She wants very much to learn how to correct typing errors but this topic is not covered until later. The current practice exercise is about a prerequisite cursor movement skill.

The learner's uncertainty about the exercise and her own specific concerns encourage her to try correcting a typing error on her own (beginning of segment 2). However, her attempt fails because she has not yet learned how to correct typing errors (it involves positioning the cursor and using backspace or delete keys), and is not able to work out how to do so from what she already has learned or can guess. Indeed, the learner ultimately inserts blank lines into her text (segment 6) and botches the exercise. These errors arise because she incorrectly tries to move the cursor using the space and return keys (segments 5 and 6, respectively) which are not appropriate on the word processing system (i.e., these keys actually change what one has typed because the system is always in insert mode, that is, spaces and line breaks are inserted into the text line wherever the cursor happens to be).

TABLE 2
Correcting a Typing Mistake Prematurely

(1) E: What are you thinking?
L: I'm trying to figure out what exercise is supposed to be for [*pause*] line advance? Supposed to be an exercise on errors, but am I supposed to be trying to make errors and then move back?

Types first three lines of file. At end of third line, types comma instead of a period. Realizes her mistake after she has already carrier returned. Wants to know how to move the cursor back to make the correction.

(2) Oh, I see, so now . . .
E: What are you thinking?
L: I made a mistake up here, now if I want to go back I guess I would. . . . *Looks in the manual for information.*
E: What are you looking at, page 3–4? [*Pause*] What is that telling you?
L: Well, I'm trying to figure out how I can go back to correct that mistake. Am I supposed to correct my mistake yet, or am I supposed to just not worry about the mistakes or . . . I'm going to try to go back.

(3) *Backspaces with the cursor at the beginning of current line in order to move cursor back to immediately preceding line. This is not allowed and incurs a beep error message.*
Oh! It didn't like that.

(4) *Presses line return, which moves the cursor to the beginning of the previous sentence with the comma which she wants to correct.*
E: OK, what did you hit?
L: I pressed line return.

A function that moves the cursor up one line

E: On page 3–4 again, what's that telling you?
L: I'm still trying to figure out . . . press and hold down line return until the cursor moves to the first line . . . that's a step but. . . . *Trying to adapt the manual instructions to her situation.*

(5) *Hits the space bar to move the cursor to the right through the line, which puts a space at the beginning of the line. Then immediately backspaces, which deletes it.*

E: You hit a space then a backspace?

Pressing backspace deletes the unintended spaces at the beginning of the line.

L: Yes. I'm just going to leave that mistake and press return.

(6) *The cursor is at the beginning of the third line, so when she presses return she inserts a blank line above it.*

Whoa!
E: OK, what are you thinking?
L: I didn't know it was going to do that. Wonder how I can bring it back.

(7) *Hits code key.*

E: What did you hit?
L: I pressed code . . .

TABLE 2—*Continued*

E: What are you thinking?

L: I'm going to try to bring it back to where it was.

E: Bring what back?

L: Bring back the line, although I'm not really sure if it matters . . . space in between there. [*points to blank line*] I press return . . .

(8) *Presses required return and inserts yet another empty line.* Don't go any further! I'm having difficulty because I'm trying to bring it back up.

Uses line advance to move cursor down to the line that needs to be "moved back up."

I'm trying to remember how to bring it back. I don't even know if it matters if it has a space in between. I can just continue typing.

Instruction Sequences Are Fragile

Learning by trying things out according to a personal agenda of needs and goals is not merely a preference. Learners who try to follow out manual instructions are often unable to do so. The instruction sequences are fragile in the sense that it is easy to get sidetracked and there is no provision in them for recovery. One example is a learner who inadvertently paginated (reformatted) a document at the beginning of an exercise on revising documents. This not only rearranged the lines in the file to make right margins even, it also stored the document away. The learner had not yet learned how to retrieve documents and the manual itself provided no recovery information for this type of error. Accordingly, she was forced to try to discover how to retrieve the document on her own. She was ultimately unable to do so and we had to intervene.

Once we restored the document to her display, she was faced with an equally staggering problem: the pagination operation had rearranged the lines of her file so that the revising instructions did not refer to the same document. An experienced user who understood reformatting could have reinterpreted the instructions and adapted them to this rearranged text. But this learner had no idea what she had done, and thus was hopelessly puzzled by the fact that the instructions seemed to be wrong.

The fragility of instruction sequences, coupled with the propensity of learners to try to recover by initiating exploratory forays, can result in problem tangles: Learners, who may not even fully understand the individual operations, have little basis for appreciating the subtle interdependence of clusters of word processor operations. They may be unable to diagnose or even recognize the problems they encounter.

Passive Learning

When learners do not, or cannot, follow directions the problems that arise can result in their losing track of what they are trying to do. It is likely, of course, that this loss of task orientation contributes to the overall failure of learning—as indicated by the trouble all learners had applying their learning experiences to the routine typing "transfer task" after training.

What is more surprising perhaps is that even when learners were able to successfully follow instruction sequences out, they still seemed to experience a loss of task orientation, as evidenced by comments like: "What did we do?" "I know I did something, but I don't know what it is!" or "I'm getting confused because I'm not actually doing anything except following these directions." For these subjects, the overall orientation toward accomplishing meaningful tasks (e.g., type a letter, print something out) has been subverted by a narrower orientation toward following out a sequence of instructions.

This is ironic, since self-study educational technology is predicated upon learning-by-doing mix of exposition and practice. The materials we used provided expository information about how to do things with the word processing system along side of step-by-step practice with those procedures. The problem seems to be that "learning by doing" means different things to manual designers and new users: following programmed exercises is "active" from the viewpoint of the designers but "passive" from the viewpoint of the learners. Unfortunately for the learners, their innocence about computers prevents them from actively learning by doing on their own with much success—at least given state-of-the-art system/training constellations.

LEARNING BY THINKING

Just as learners take the initiative to try things on their own, so also are they active in trying to make sense of their experience with the system. Learning passively by rote assimilation of information is atypical. Rather, learners actively try to develop hypotheses about why it operates the way it does. These quests after meaning can be triggered by new and salient facts. They can be forced by discrepancies between what is expected and what actually happens. They can be structured by the learner's personal agenda of goals and queries, referred to as new problems arise. And they can be resolved by analysis into elementary procedures in the context of some word processing goal. In each case, learners' lack of knowledge about word processing makes it difficult for them to reason out coherent solutions that accurately represent the objective operation of the word processor.

Constructing Interpretations

Because learners have no basis for recognizing and ruling out irrelevant connections, their interpretations of word processing systems are often influenced by spurious connections between what they think they need and what they perceive. The protocol example in Table 3 provides an example.

A learner is trying to find some way of using a "create document" command to retrieve a document (beginning of segment 2). The command is initiated by filling in parameters of a create document task menu. However, "create document" won't retrieve a document that already exists. Hence, when she fills in the name of the stored document (toward the end of segment 2), she incurs an error. Because the learner does not know what is relevant to getting a document back, she grasps for other possibilities (segments 6–8). The word "number" in a menu parameter called "document charge number" suggests to her that typing in page numbers will retrieve those pages for her. The word "originator" in the parameter called "document originator" suggests that typing in her operator name might somehow tell the system she wants *her* document. Neither parameter has anything to do with document retrieval, and both merely evoke further errors.

Learners are often faced with situations where they perceive a need to interpret some fact or observation in order to make sense of it. Often these interpretations are not necessary for understanding how to do something, but simply reflect their desire to *make sense of their experience*. For example, more than one learner was troubled by an inconsistency in the functions which are used to move the cursor around the screen. One set of functions is used to move the cursor within a menu display, while another set is used to move it around a typing display. The learner reconciled herself to the distinction by reasoning that a cursor movement function which moved the cursor from one line of a document file to another would not work in a menu display, because there could be more than one field on a line of a menu and she decided the function would not be able to move from one field to another *within* a line of a menu. In both cases, the learners could have simply accepted the way things were accomplished. Instead, they tried to construct rationales for why things were done the way they were.

In other cases, learners are faced with the need to *interpret discrepancies* between what happens to them, and what they think should happen. Because they lack knowledge about how the system operates, these ex-

TABLE 3
Not Knowing What Is Relevant to Retrieving a Document

(1) L: Did that make any sense to you? It didn't to me. I don't know why we had to do that, that was ridiculous. Now, let's see how do we get the pages back, that's going to be fun.

(2) *In a selection menu listing things users can do, learner presses enter to bring up a menu which when filled in and entered will create a new document file.*

You know what it's probably going to do now, it's going to tell me that there is a memo in there already. [*She is correct: i.e., she has already created a document named "Memo"; she wants to retrieve, not create this*]

Looks in manual for help.

E: What are you thinking?

L: Well, they were supposed to tell me anyway how to get these pages that I just lost.

In the create document menu she types the name of the document that she wants to get back. A message, "document name already exists," appears.

Document name already exists, I knew that was going to happen because I just typed it in.

(3) *Presses some other keys.*

I don't know how to do this.

E: What did you just press there?
L: Reset.
E: And before that?
L: This again, cancel [*which has no effect when the selection menu for tasks like create or revise is on the display, because there is nothing to cancel*].

Request. *Presses request, cursor in now on "request line."*

I'm just going to type "memo1" and press page return to see if it comes up [*"memo1" is the name of the document*]

Presses request. Looks in manual for help.

E: What are you looking up?

(4) L: I might have to look it up in one of these documents [*refering to tasks listed on the menu. The task or command she needs is listed there*]

I don't remember which one it was. [*Because she has not learned it yet*] Create a document . . . I can't use that.

(5) E: Why don't you think about it a little bit more, and if you still can't figure it out we'll help.

Reads the manual. Presses an invalid key.

L: Invalid key . . .
E: What were you looking for there?
L: Some solution to this problem. . . . We're at a stand still here. . . .

(6) E: You pressed reset a couple of times there?

Without answering presses request, types "cre," and the create document menu appears.

TABLE 3—*Continued*

---

(7) E: What are you thinking?

L: Would this be, "document charge number," would that be the page number? [*It has nothing to do with page number—she is still in effect trying to create a document*] Maybe I have to look it up by the page number.

E: What are you thinking?

L: Last time I wrote the document name in there it didn't take it the way I wanted it to take it.

Advances cursor to the "document charge number" parameter.

E: What did you conclude from that, what you just read there?

L: No, No, I didn't read this. [*i.e., she had not noticed the "document charge number" parameter before in the menu for creating a new document*]

Deletes the "001" charge number and types in "0002" page number but overflows the menu field limits and gets an error message to this effect.

Doesn't fit. [*Presses reset to acknowledge error*]

E: Press reset there?

Deletes the "0002."

(8) L: [*Reads*] "document originator," would that be me?

Advances cursor to yet another menu parameter, this one called "document originator." This is where one indicates whose document one is typing, e.g., in an office, one's boss.

She types "learner1", presses enter to enter the create document menu; but because she has not specified a name for this "new" document, she incurs an error "Parameter omitted or not valid." The actions are irrelevant to retrieving a document anyway.

(9) E: What are you thinking?

Learner realizes that she has "paginated" document prematurely rather than after she has completed the exercise she was about to begin. That they should explain that it was going to be done later on. [*i.e., paginate or cause the rearranging of lines of text to right justify*] Cause they said before you do all of this you have to paginate. You will also, oh, no before printing it. So, in other words I assumed that I was supposed to do it right at the beginning, but where as I like wiped myself out. [*In fact, she has not deleted anything—she simply does not know how to retrieve the document file*] But it was done much later on, down here.

---

planations often are not accurate representations of what is really happening. In one case, a learner tried to decide if a "file" command has stored a document file away. It was not stored because the command was entered in a text input mode where all typed strings are interpreted as text, and not executed as commands. But she assumed that the file had been stored, and adduced evidence to confirm this premise. Not knowing what is really relevant, the learner found two features of the screen that were superficially similar to what one might hypothesize if the file command had been successful, and the learner concluded the document had been stored. For example, at one point she notices a status message "input mode 1 file" which indicates that she is in the text input mode. However, the word "file" matched her file command, and this was enough to suggest some kind of feedback that her "file" (as in store document) command had worked.

Table 4 illustrates a slightly different kind of reasoning in which a learner copes with a number of problems as she tried to sign on to a command-based word processing system we studied. For example, in segment 1 she is trying to decide where to enter the password. The correct point is the command line on the bottom of the display (where other commands have been entered). However, the explanation of signing on that she is following is not explicit about this, and the learner fills in by trying to type the password at the top of the display next to the "password:" prompt (perhaps she thought of this as a blank line of a form that one might fill out). This area of the display is a protected area and her attempt to type locks the keyboard and cursor. But in segment 3 she hypothesizes that inability to move the cursor at that point means the cursor is already in the right location to type! Finally, in segment 4 the learner seems to conclude that the lack of response to her password is the result of heavy load on the system!

How can we characterize these reasoning processes? In some cases, reasoning appears to consist in *adducing* factual support to a premise the learner would like to hold as true. Table 4 shows that the learner began with the hypothesis that she had stored the document file away, and sought evidence to confirm that this was the case. Her adduction here was incorrect because she did not know which facts were relevant to verifying the premise. She was persuaded that the superficially relevant word "file" in the status message signaled successful storing of the file.

In other cases, reasoning appears to consist in *abducing* (Peirce, 1958) a hypothesis when it, together with other assumptions the learner may already hold, is consistent with some fact or observation. An example in Table 4 (segment 1) is the learner's hypothesis the password is entered like an item in a form, that is, after the password prompt with the colon (rather than on the command line like other entries into the system). Other examples of abduction and a fuller discussion can be found in Lewis and Mack (1982).

Abductions and adductions are often incomplete and partial: people do not test them against all potentially relevant data. In particular, people tend to reason only from confirming evidence—and to overlook potentially disconfirming evidence (see Nisbett & Ross, 1980). As a consequence, learners are able to construct and verify ad hoc interpretations of what happens which may misrepresent the situation they are actually in.

Nevertheless, adductive and abductive reasoning processes are important in that they afford net growth in knowledge without requir-

TABLE 4
Abductive Inferences in Signing On to the Text-Processing System

(1) L: OK, let's try it again. [*i.e., type password*] I'm on the wrong keys. Password, now type in your password. Now I can only type by going up correct? It's not telling me that. I'm assuming I can take the cursor up there.

I.e., learner does not know where to put cursor. In fact, it is already on the command line where the password should be entered. But the learner decides to move the cursor up the display right after the password prompt at the top of the screen.

(2) *After moving the cursor up the screen, she tries to type password, but this is a protected area of display which causes the keyboard to lock until a reset key is pressed. An "input inhibited" light also comes on.*

I expected it to come out. I expected to see the word. [*i.e., password*] but it's . . .

E: What happened?
L: It probably fed it and I made, umm, contact without showing it here, because it is a password and not everybody knows it generally. So it's sort of a secret between me and the computer.

Rereads description of password.

"The password will not print what you typed. If you are correct the display will blink and in a moment a few lines of greeting messages . . ." and evidently it's not right.

E: OK, what makes you think it's not right?
L: Cause it's not blinking. OK, do I? Ok, let's see the display will, at this point the display will . . . the word will appear, um.

The keyboard is locked, the cursor is in a protected area to the right of the password prompt.

(3) I assume maybe I should space one [*after the password*] Would that maybe [*work*]?

Tries moving cursor one space to the right of colon using space bar, cursor movement blocked because of locked keyboard condition.

Well it doesn't even space so I'm in my right position.

Tries typing password again

E: Ok, you just typed it again.
L: Shouldn't I hit enter?

(4) *Notices the input inhibited light on the display.* What do you think that means?

E: What were you pointing at?
L: The input inhibited [*light on the display*]. Is that possibly why that's not accepting that at the moment?

[*After brief digression*] I've seen that all too often. *She had had some experience using terminals for data entry.*

E: Well, what do you think that means?
L: So it's being held up temporarily. I don't think there's anything you can do at the moment until the light goes off. Oh it might be accepting my, it could be accepting my . . . what I put in and it will go off as soon as it's um . . . When you put information on there now maybe it . . . the input, the fact that you're putting in . . .
The learner is unable to develop any hypothesis as to why the system will not respond or accept input. The experimenters later were forced to intervene.

ing extensive prerequisite knowledge. In contrast, both classical methods of reasoning—deductive and induction—impose this requirement. Deduction requires an extensive knowledge base of principles from which univocal predictions can be derived. Induction requires an extensive and systematic empirical data base—and in any case cannot guarantee the validity of its conclusions. Neither of these two methods would be appropriate for our learners. This is because the information they have available about the system is typically too impoverished to allow them to build up representations by systematic induction and deduction, even assuming that people are sophisticated and motivated enough to do this. Abduction and adduction provide means of generating and supporting hypotheses from limited information.

Setting Goals and Solving Problems

Beyond merely trying to interpret experiences, learners often set goals which they actively pursue by trying to solve problems. They are hampered in this by not knowing the appropriate problem space, or domain of possible actions and interpretations relevant to accomplishing goals and addressing queries. Accordingly, their strategies are often local and fragmentary; they have difficulty integrating information or other experiences, and in formulating their concerns in ways that map transparently onto system functions. Learners appear to construct a personal agenda of goals and queries as they go along. As situations arise, this agenda is referred to opportunistically: fragmentary aspects of the local situation are assimilated to some standing goal or query.

Table 3 provides an example of the local and fragmentary character of setting goals and solving problems. Recall that the learner in this example is trying to retrieve a document she has inadvertently stored, but does not yet know how to do this. Segments 2 and 4 suggest that the learner has a relatively accurate general idea of what her problem is and what goal she should pursue: retrieve a document. In segment 4 she even tumbles to a relatively good strategy for solving the problem, namely, look for a command listed on the selection menu which could retrieve her document. Unfortunately, the correct command, while listed there in fact, evidently does not have a suggestive name and the learner drops that strategy without trying anything out.

Instead, beginning in segment 6, the learner tries to exploit a create document task with which she is already quite familiar. While the learner again reveals partial insight into her the adequacy of her strategy by suggesting (correctly) the she will incur an error using the create task (users cannot "recreate" existing files), she nevertheless incurs an error when she tries to retrieve her document using this command.

In light of these experiences, segments 7 and 8 are especially interesting. Despite her interpretation of the inappropriateness of the "create document" task for retrieving the document (segments 2 and 4), the learner returns to this menu and hypothesizes that other parameters of the task might be relevant to getting her document back. As we discussed in the preceding subsection, the learner does not know what is relevant to her goal, and so can entertain even specious hypotheses about what might be relevant based on very superficial resemblances. For example, the learner fills in the page number in a menu field labeled "document charge number" because the word "number" suggests page numbering.

Table 2 describes another example of how difficult it is for learners to formulate goals, and find information relevant to solving them. In segment 8 the learner tried to delete a blank line she had inadvertently inserted between two sentences. However, she did not describe her problem in this way, but thought of it as one of bringing the bottom sentence back under the first one (segment 6). The relevance problem arises during her search for some way to "bring the sentence back": the label "Required Return" on a carrier return key seemed to be relevant to this problem. In fact, the connection is specious. The Required Return function is just another version of the return key and resulted in the learner inserting yet another blank line.

When learners cannot solve a problem or resolve some query immediately, it may be saved away on a task agenda—with the hope that an answer will be found later. The goals and queries on this task agenda are typically addressed opportunistically rather than through systematic problem solving. (See Hayes-Roth and Hayes-Roth, 1979, for related views of planning behavior.) An example is the learner who initially misinterpreted why the word processing system did not let her sign on (i.e., in segment 6 of Table 1, the learner thinks she must be in a "Test Mode" to work on the system). Later (segment 7), when she dealt with a similar problem situation it occurred to her that her current (correct) interpretation was probably applicable to her earlier difficulty. Not only did this learner reconsider her earlier erroneous interpretation, she actually tried (segment 8) to recreate the original problem situation.

The learner in Table 2 who tried to correct a typing error before she had learned how to do so had been anxious for some time to discover how to correct typos. In another case (not shown in tables), a learner had unintentionally accumulated a number of special characters ahead of the cursor. The system is always in insert state so this material was pushed along as she typed. Of course, she wanted to eliminate this unsightly mess but did not know how (the delete key would have worked). It was only much later while she was learning how to delete another kind of special character that she realized the same operation might solve her earlier problem: indeed, when she next had the opportunity, she performed the appropriate deletion.

Consolidating Procedures

The word processing domain is highly procedural. Learners must analyze basic procedures in the context of various word processing goals. This is not a matter of rote learning and passive assimilation; even the simplest procedures like word deletion or replacing a letter reveals the difficult task of identifying relevant elements of these procedures and trying to integrate them into a smooth operation. Learners' understanding of procedures works by successive approximations. Learners form general schemata or rules for those elements relevant to a procedure. These general schemata are then filled out in the course of interacting with the word processing system.

Table 5 shows an example of a learner who is trying to delete the underscoring for the three words, ". . . will not change." The underscoring itself looks normal but it is marked by a special underscore character which is a block of reversed video superimposed on the space immediately following the word. To delete it, the user must position the cursor "under" the character in such a way that a message "word underscore" appears on an information line at the top of the display. (It is not enough that the cursor is spatially under the reversed video block.)

The learner's attempts to coordinate her initial understanding of this delete operation demonstrate the *revelatory nature* of trying actions out: a sequence of actions and events is consolidated into a procedural concept. The learner has the correct general idea of positioning the cursor under a character and then using a delete function to delete it (evidenced in segments 1 and 2), but does not incorporate the critical feedback from the information line at the top of the screen. The procedure has been incompletely instantiated. Even though the learner has read all this, and as the example shows will read about it all again (e.g., segment 5), its significance only becomes apparent through careful and detailed attempts (segment 6) to actually integrate and exercise the procedure.

One obstacle to integrating the elements of procedures, again, is the learner's innocence. In Table 5 (segments 7 through 11), the learner tries to replace a small "t" with a capital "T" in the word "the." The learner deletes the "t" correctly, but then formulates her goal of replacing it with a "T" to be one of replacing the space at the front of the string "_he" rather than simply inserting the "T" at the beginning of the string. Of course, the "T" is in effect inserted in the existing string which now includes a space, producing the string "T_he." In segments 9 through 11 she struggles to understand this procedure by tying the

TABLE 5
Deleting a Word Underscore and Replacing a Letter

(1) *The learner is trying to delete the underscore from the second of three words "will not change." She is not locating the cursor in a way that identifies the special underscore character*

L: What happened? The line [*underscore line*] didn't disappear, now what did I do? Was I over too many spaces? I could have been.

(2) Let me try again. I can replace the other one anyway. *Presses character delete, deleting the "c" of the third word "change." Not clear why she did this.*

What's happening?

Positions cursor immediately after the third word and "under" the underscore instruction, so that the character delete operation correctly deletes the underscore from that word.

It worked on that one, I don't understand. *Here is the problem: Underscore is represented by a special character symbol. To delete, the cursor must be positioned "under" this symbol and then deleted as any other character. The learner did not enter underscore instruction at the correct location for the second word but did so for the third. Therefore her action of locating cursor "under" the underscore character works as per instructions for the third, but not the second word. The learner does not understand what the instructions really mean or she would be able to recognize the discrepancy, and modify the instructions for the second word.*

(3) *Presses word delete key, and deletes the third and fourth words.*

E: What did you just press there?
L: I just pressed word delete.

Moves cursor to the next word and deletes it.

Oh, wait a minute, that keeps moving all of them. [*i.e., the line adjusts after material has been deleted*]

Looks in manual for information Checking out something here under [*the topic of*] word delete.

E: What is it that you're looking for?
L: I'm just trying to see . . .

(4) *"Not" is still underlined and every word in the line after "not" has been deleted.*

I got a little paranoid when I saw everything moving backward, I thought "wait a minute." Now, I realized I wanted to erase the whole thing anyway and retype it and then. . . . Because I lost a space in there and I couldn't get rid of the line under "not."

E: What were you looking for when you looked under [*the topic of*] word delete?
L: Yes, for some reason I was afraid. I saw this moving over and I saw this moving over and I was afraid that I was going to erase. I don't know I'd . . . I'd forgotten that everything was going to move over anyway as you erase, as you delete. So, I didn't want to go on to the next word, although actually I may have to erase that also. OK, I'll see, I'll see if I can type this, retype this line . . .

Types in the rest of the sentence, after underscored "not."

Oh, that's OK, alright.

(5) *Moves the cursor back to the space before the underscored word "not."*

(*continued*)

TABLE 5—*Continued*

E: What are you looking for now?

L: OK, I'm just looking to see what it says here again. [*I want to*] start again with the underscore . . . deleting the underscore. . . . Yes, no I'm just going over what I did before.

Still trying to delete the underscore line under "not," places the cursor under the space after the "t" and deletes a space.

Oh, what did I do?

(6) *Moves the cursor under the "t" in "not," then moves the cursor to the space after the "t."* I see, OK, I didn't realize that before. I didn't realize it [*the cursor*] has to pass through . . . the underscore. It has to pass through the underscore from what I understand.

E: How did you find that?

L: Um, under the, under the delete underscore instructions.

When she moves the cursor under the "t," the message "word under" appears on the screen.

I did that automatically before. It was supposed to be under the last letter of the underscore. *Presses a delete function with the cursor "under" the underscore symbol; the underscore is deleted.*

(7) *Learner moves on to another instruction for revising, this one involving replacing of a "t" with a "T" in the word "the" which is the first word of a newly revised sentence.*

(8) *She deletes the "t" in "the" to produce the string "he" with the cursor positioned under the "h." But before she tries to type the "T" she moves the cursor to the left, back under the space before the "he." Thus when she types "T" she actually inserts it in front of the space to produce "T he"; i.e., with a gap.*

E: What are you thinking?

L: OK, um, when I deleted the "t" and typed the capital "T," I created a space. Why is it doing this?

(9) *Looks in manual for help.*

E: Where are you reading now?

L: OK, I'm reading under character delete, I'm just . . . I'm trying to find out why when I retyped another character in place of the one I just deleted . . . um, I'm just wondering, there's another space after that now separating "T" and "h."

(10) *Places the cursor in the space between "T" and "h" and inserts another "T." Perhaps she is convincing herself that this really happened. Then she turns back to the manual section on deleting material.*

It says they move together.

E: Was that a particular thing you just said?

L: OK, it says, the instructions say that when you delete something the letters will move together to fill it in. OK, now that. . . . I don't understand why there's a space there as soon as I type the other one.

(11) [*Sometime later learner speculates*] When you delete, you take out. Do you add when you put in?

"addition" of text to her earlier experience deleting it. As she puts it in segment 11: "When you delete, you take out. Do you add when you put in?" Her real problem though is not recognizing that an important element of this procedure is the fact that the word processing system is always in insert mode.

A second problem, we have also seen before, that confronts this process of integrating basic procedures is that of problem tangling. In Table 5, for example, the learner had incorrectly underscored the second word of the triple (i.e., in the triple "will not change," the underscore character for "not" was originally entered at the last letter of the word, not under the interword spacing). Thus, while the learner was able to delete the underscoring for the third word "change" (segment 2), the same procedure mysteriously does not seem to work for the second word "not." The problem could be solved either by recognizing the earlier error, or by recognizing the correct feedback for locating the underscore character. Since the learner cannot untangle these problems, she cannot properly execute the procedure, which remains mysterious to her.

These examples suggest that even the simplest operations present learners with serious challenges. They must translate expository descriptions of operations into actions. They must coordinate multiple sources of information (on display, in manual, and from memory). And it is clear that the significance of the various elements of procedures (what they accomplish, what prerequisites are needed, what are relevant outcomes, etc.) only become tangible to learners in the context of their actually interacting with the system.

LEARNING BY KNOWING

Learning by reasoning and problem solving presupposes a well-defined "problem space": knowledge of what is relevant to particular goals, of how to constrain interpretations and goal-related actions, etc. Our learners understood the problem space of the conventional office, but not that of the "electronic office"—in particular, they did not understand the problem space of word processing. Nevertheless, they spontaneously referred to substantive prior knowledge from the former domain in order to understand the latter. We will refer to this as "learning by knowing"; it has often been called "metaphor" (Bott, 1979; Carroll & Thomas, 1982; Rumelhart & Ortony, 1977).

To this point, we have argued that a new user of a word processor relies without instruction on active exploration and ad hoc reasoning as learning strategies. However, not all possibilities are explored and not all hypotheses that could be reached are reached. What constrains these strategies is a sense of what could be appropriate—and this devolves from prior knowledge on the part of the learner: knowledge about devices "like" word processors (e.g., typewriters), knowledge about office routine and work in general, even knowledge culled from interacting with the word processor up to that point in time.

The Typewriter Metaphor

Office personnel learning to use a word processor typically have an extensive and fairly coherent body of knowledge regarding typewriters to which they can refer. Indeed, our learners were not able to resist referring to their prior knowledge about typewriters as a basis for interpreting and predicting experience with word processors.

One of our learners came to a halt as she read an instruction in the manual which said "Backspace to erase." It seemed that she could not interpret this instruction for, as she pointed out, backspace does not erase anything. She had irresistibly availed herself of her knowledge of how backspacing works on a typewriter, unable to even consider that this knowledge might be inappropriate for the present case. The incident had a sad ending, too. In wrestling with this problem, she accidentally incurred a "reset" condition, repositioning the cursor at the beginning of the word. Hence, when she finally elected to plunge ahead and "backspace to erase," it in fact did not work.

For a second example, recall the learner in Table 2 who was trying to move the cursor through a line of text to correct a typing mistake. Without any hesitation, she presses the space bar. She is surprised to see that this inserts a space into her text. Now, just as automatically, she presses the backspace key. This deletes the space. What is interesting in this example is first, that the space bar is expected to advance the cursor without inserting spaces, as it would do on a typewriter. Second, it is notable that when it fails to behave as anticipated, backspace is employed as an inverse operation—again just as it is on the typewriter. Neither of the expectations is thought out, rather both are immediately available in virtue of the learner's prior knowledge of typewriters. Fortunately, although the first expectation is thwarted, the second obtains: the space bar fails to provide a vehicle for cursor movement through text, but backspace succeeds in inverting the operation (including the undesired side effect).

To continue with this example (segment 5), the learner decides to abandon the goal of correcting her typing error (possibly because she cannot manage to get the cursor to the site of the error). She now wishes to advance the cursor and continue entering text. Again, with no deliberation, she keypresses the carriage return (segment 6): a blank line is inserted into the text. "Whoa!" The learner is surprised by this. Like the side effect of the space bar, it is discrepant with her expectations based on typewriters (where carriage return advances the typing point but does not insert lines)—and worse, there is no obvious predictable inverse operation available to her.

Expectations about Work

Our learners were experienced with conventional office work: typing letters, filing, etc. Their knowledge about how these routine tasks are organized in the office creates expectations in them about how analogous tasks ought to be performed in the "office of the future" (as represented by the word processor in our laboratory). Thus, one response to revising a letter task is to retype. This is striking since it is the capability of the word processor to store and retrieve documents—for revision, among other things—that is its fundamental advance over previous office technologies.

Another example of mismatched expectations is the correspondence between the typing page on the word processor and what learners expect a typing page to be from typewriting. The system ostensibly frees users from worrying about margins and formatting by doing these things automatically during printing, and having default settings. But new users can be troubled by these labor-saving features: the differences between a familiar piece of paper in a typewriter and the typing display raise many questions that are not easily answered early in learning. Learners wonder where the top and bottom, left and right margins are located on the display. They wonder how the length of the typing display corresponds to the length of a printed page (this is a relationship automatically controlled by the pagination feature). And they wonder whether error and status messages which appeared on the typing display would be part of their final product.

The learner's expectations about work transcend knowledge of how office procedures are carried out. They include motivational expectations about work-related learning and sociological expectations about power and control in work-related settings. These expectations can differ remarkably, as evidenced by comparing our learners to a college teacher learning to use a programmable calculator (Carroll & Lasher, 1981). This person was in fact, suffering many of the problems we have described above. In one particular incident, she was unable to make sense of several passages in the training manual.

Now in analogous situations our typist-learners typically concluded that the fault was their own, that they were too stupid to understand properly. However, our Ph.D. learner came to an entirely different conclusion: she decided that the manual contained misprints or other errors and could not, therefore, be used. She discarded it at that point—*never* considering the possibility that she was too stupid to understand in general, or even that the specific trouble was in part due to her own miscoordination of information. We believe that this manifest difference in attitudes is partly because the typist subjects identify with social powerlessness and hence expect to be victimized in work-related undertakings. However, we are basing this on a single comparison subject, and our interpretation must be weighed accordingly.

A final point we will simply make now and then return to later is that our learners seemed to expect a dichotomy between work-related learning and play. Of course work and play *are* different classes of activities for most people, but in learning it can be useful to adopt some aspects of a play orientation. Just as a causal observation, it seems that the capacity to shift gears a little and to treat work as play is characteristic of successful adult learners and problem solvers. The reluctance of our learners to do this (which was something the manual encouraged) may constitute a fairly abstract orientational learning obstacle.

Integrating Bits and Pieces

As a learning experience progresses, the learner is acquiring and organizing new bits of knowledge. The ultimate goal—and the final measure of success in the learning situation—is that of assembling these pieces into a coherent fabric, an understanding of the word processor. Along the way, any prior bit of knowledge is available for use as a basis for expectations concerning successive interactions with the system. Learners expect functional consistency from operations with similar names. One system we studied contained inconsistencies of this sort. Thus, to delete a word, one positions the cursor under the word's initial character and keypresses a word delete. However, to underscore a word, one positions the cursor under the final character of a word and keypresses the word underscore command.

This inconsistency caused one learner to misexecute one and then the other of these two operations in a dismal cycle of negative transfer as illustrated in Table 6. In two earlier episodes (not described in the table) the learner had followed explicit instructions for underscoring and deleting, and understood both. The episodes shown in segments 1–7 reveal that the learner eventually formed correct rules to cover both cases. For example, after learning how to underscore and delete on the first

TABLE 6
Learning Where to Position the Cursor for Deleting, Underscoring, and Backspacing

---

(1) *On Day 3 learner is about to underscore the word "solutions" but has positioned the cursor under the first letter, rather than at the correct position at the end of the word*

E: What are you doing?
L: I'm trying to code word underline [*i.e., To underline a user must press a word underline function in conjunction with a code function*]

Today I remembered which key it is [*she had trouble finding the word underscore key the day before*] but evidentally I didn't hit it in the right sequence.

Not clear what reference is, but in her first experience with the function the day before, she had pressed the code key after the word underscore key, which is the wrong sequence for underscoring—but her current problem derives from having a wrong cursor position.

Why didn't it underline? Oh, I should have . . .

E: Can you talk aloud about what you are thinking?
(2) L: Yes. Ah, I'm thinking, I'm looking at the word "solutions" on page 3–19. I wanted to underline it. I brought the cursor back under the "s" [*first "s" of the word*] before I hit the code underline. And I think that the code underline should have been hit at the end of the word. [*She is correct: i.e., the cursor should have been at end of word, not beginning.*]
E: Why do you think that?
L: Because it only underlined one thing [*the first "s" of "solutions"*]. I think.

She is correct. She then moves the cursor to the end of word and correctly underscores

(3) *On Day 2, learner is trying to reconstruct the correct position of the cursor in order to delete a word. She had learned the day before that the correct position is at the beginning of the word*

L: They have a code word delete. [*i.e., reference to the necessity of holding a code key while pressing a key labeled "word delete"*]
E: They what?
L: Code word delete. Now what I'm trying to figure out . . . in these instructions they didn't tell me where the cursor should be when I do the deletion. I think probably right after the last character of the word. So we'll go see.

We're going to get the cursor down to the word we want to work on. So we are going to line advance twice. [*i.e., press function to advance the cursor down a line*] And the word we want to work on is "have." So we are going to word advance to that [*i.e., use a function which advances the cursor to the next word within a line*]

Cursor positioned at beginning of "have." Now I think probably that I want to be at the end of it to eliminate it. *Uses character advance key to position cursor at end of "have."*

Okay as though I had just typed it and now they have a two down [*numbered manual instruction*] they want me to code word delete. So. Find "word." I want to be sure I'm watching [*the display*] so that I can see what happens.

(4) *Pressed code plus word delete and deletes space between "have" and next word so they are concatenated*

Oh no!
E: What happened?

(*continued*)

TABLE 6—*Continued*

L: I lost the space. So I have to put the space back in.
Presses backspace which deletes the "e"
Oh no! Now I lost the "e"!
E: What . . .?
L: Terrible!
E: What did you press to [*try to*] put the space back in?
L: I pressed the backspace [*which deletes*].

(5) Okay, now let's see. I tell you what I'm going to do. I going to press the "e" which I lost. [*Inserts "e" to restore "have"*] It moves it over [*i.e., inserts*]. Then I'll press a space which I lost *Presses space bar to insert space between "have" and next word.*

I'm going to take the cursor and character return it to the "e" [*i.e., uses function to return cursor one position left; in this case under "e" of "have"*] and then I'm going to code word delete again and see what happens there.

Ah, my whole problem here is. . . . *Positions cursor under "e" and presses code followed by word delete which again deletes the "e." Repeats to now delete the interword space.*

No, that wasn't the answer either.

E: What did you do?
L: I pressed code word delete and all I lost was a letter [*i.e., "e"*].
E: That's not what you expected?
L: No. I expected to lose [*delete*] the whole word "have."
E: What makes you think that?
L: Well, I thought it would take the whole word out.

(6) Now maybe I coded word delete in the wrong place. So let's put the "have" back in, the "have" and the space. Um. Is the space in? I'm going to go all the way back to the beginning of "have" with the cursor.

Repositions cursor under "h" of "have." And at the beginning of "have" I'm going to put code word delete and see if it takes the whole word "have" out.

Presses code and word delete which deletes "have". Yeah. That was the proper way. So when you use that you have to be at the beginning of the word you want to delete.

(7) E: Can you think of any particular reason why you thought that? That it had to be at the end of a word rather than at the beginning?

L: Because you can't delete it unless its already there. And if you were typing on you might decide you wanted to delete something right after you typed it.
E: I see.
L: And you would do it at the end. But I see that wouldn't work. You would have to get the cursor back to the beginning of the word if you wanted to delete the whole word.
E: Does that make sense to you now?
L: Yes. It's all right. If that's the way it works, that's fine. I still don't know why I was told to code line delete at the other one instead of coding word delete. And I'm very grateful but I don't know why I didn't lose one of those lines.

(8) *On Day 4, learner is reconstructing how to underscore a word. She tries to position the cursor at the beginning of the word, rather than correctly at the end as she learned on at least two previous occasions.*

TABLE 6—*Continued*

L: Just sitting in the office I would just take a chance that that would underline.

Pressing word underscore with the cursor under the first letter of the word underscores that letter only, and superimposes a block of reversed video to mark the underscore character.

If it didn't I would have to go to the book and find out.

(9) E: Well, this is just a question. I think we are beginning to wrap up. In the past when you used the word underline did it actually put underlining on the screen?
L: I can't remember today [*in fact, it had underlined completely*]
E: You can't remember whether it does or doesn't?
L: Whether it did or did not. But I'm not particularly worried, because the rectangle means there is an instruction at that point and the only instruction I put in was word underline.
E: Right
L: And I think I put it in at the right place. I think to give an instruction you have to be at the first character of that word. Like in other words to delete the word I found out you don't delete it at the end. You can't type delete just after you typed the word.

You have to get the cursor under the first character of the word and then type word delete to take out the word.

(10) E: Can I ask you a question about that? For awhile we noted that when you wanted to delete a word or a line you put the cursor at the end of that respective unit, whether word or line
L: Yeah, I didn't know how to do it.
E: Why did you do that?
L: Because I figured that it was the last thing entered and that it might just erase the last thing entered. Not a very . . . oh, also I think on a line delete I did that, too. I don't remember what happened whether it took out the next line or it didn't because the carriage return was in there. I really don't know what happened.

She is referring to an earlier experience in which she tried to delete the last part of a line with the cursor incorrectly located at the end of that line. It didn't work but was sufficiently ambiguous that this may not have impressed her.

---

day, segments 1 and 2 reveal that on the third day the learner erroneously tried to underscore with cursor at the *beginning* of the word ("solutions"). Why did she assume this after learning the correct procedure earlier? The protocols are not clear in this segment, but the same problem arose on the fourth day and in this case the learner's explanation suggests a generalization based on deleting words.

This second instance of trying to underscore with the cursor at the beginning (rather than at the end) of a word is shown in segments 8–10. When we finally asked her why she did this, she suggested: "I think to give an instruction you have to be at the first character of that word. Like, in other words, to delete the word I found out you don't delete it at the end [i.e., with cursor at the end of the word] . . . You have to get the cursor under the first character of the word and then type word delete, i.e., press the word delete function, to take out the word". This generalization about deleting, was hard-won in its own right as segments 3–7 reveal. The learner spent considerable time working out the fact that she must put the cursor at the beginning of the word she wants to delete, rather than at the end.

A similar inconsistency in this system involved cursor movement. In a menu, the cursor is advanced from one line to the next by pressing a variable advance key. However, in text the cursor is properly advanced from one line to the next by means of a *line* advance key. Learners tended not to acknowledge this distinction and to flop back and forth depending on whether they had more recently been on a menu or a typing display. Of course, this reliance on immediately prior experience led to numerous errors.

The problem of prior knowledge is not a simple one. On the one hand, we might say that learners will rely on their prior knowledge, and that developers of new technologies ought to take this fact of human cognitive learning into account in designing systems and in designing the educational technologies that accompany and introduce them. However, this is paradoxical: can we reasonably expect that new technologies can be constrained by old ones in this way?

More specifically, should long-term innovations in word processors be constrained by nonoptimal properties that typewriters merely happen to have as an accident of their history? Can we even find suitable metaphors for presenting new technologies to learners? Again, to be specific, should a radical innovation in word processing be eschewed merely because there is no apparent way to couch it as an extended super-typewriter?

DESIGNING FOR EASE OF LEARNING

Learning, as we have tried to suggest, is an active process. It is inescapably directed by the learner. This fact is not altered when educational technologies—like self-instruction manuals (and the systems to which they refer)—are designed to place the learner in a relatively passive role. Rather, the learning experience becomes chaotic, frustrating, and inefficient.

Cognitive science, by focusing attention on ecologically valid learning situations of nontrivial complexity, can address the theoretical and practical issues underlying the current misfit of state-of-the-art learning technology to human learners. We close this discussion by sketching three projects this might encompass.

Conceptual Framework

Learners do not "absorb" knowledge: they create, explore, and integrate knowledge. New word processor technologies, and the educational technologies that support them, must take this picture of learning seriously. Prior knowledge, for example knowledge of typewriters and routine office tasks, seems to be almost accidentally relevant—sometimes transfer is positive and sometimes negative. And this is not to presume that the issues are simple; it is merely to observe that they have not been addressed seriously to date.

We should stress that we are not urging that metaphors be explicitly and discursively presented to new users (cf. Halasz & Moran, 1982). If a metaphor must be explicitly taught to a new user, it is contributing to the amount of material that must be learned instead of relieving this burden. The best metaphor is obviously one that is implicitly and automatically suggested to the user merely by the appearance and behavior of a system. Such a metaphor maximizes the potential savings in learning effort. The well-known Query-By-Example system (Zloof, 1977) did just this by suggesting the metaphor of paper tables to users of a data base query system. The Smalltalk programming language also attempted to make use of familiar metaphors (Ingalls, 1981).

Of course, not all aspects of all systems can be as easily and elegantly fitted to metaphors. Indeed, our experience in studying the learning process in this area, inclines us to stress how easy it is to *underestimate* the subtlety, the novelty, and the complexity of the design features of computer systems. What metaphor might cover the "data stream" concept, that is the fact that the system's internal representation of objects like the "typing page" are actually linear strings of symbols? Failure to appreciate the data stream concept may indeed underlie many of the "quirks" our learners struggled with regarding insert mode, line return characters, and the like (e.g., a "blank line" on the "typing page" is represented internally as a single character). Suggesting a misleading or inconsistent metaphor through a feature of a user interface may create more learning problems than it mitigates.

Must new technology be burdened with "metaphoric compatibility" constraints? In designing the word processors of the future, will we always have to try to pretend that they are typewriters for the sake of our innocent new users? It seems to us that this is not the way to put things. For example, one can imagine the following possibility: a new office system is created—it is indeed not merely just another super typewriter (we leave open what it might be). How can it presented to new users to reap the benefits of prior knowledge and avoid the pitfalls?

A programmatic solution would be to envision a series of user interfaces; the first very much like a typewriter (in effect concealing from the user much of the innovative function that the system has to offer). The second user interface, is richer—less strictly tied to typewriter concepts—and for users who have progressed to mastery of the initial interface. The program should be clear. No one today thinks of a typewriter as a super pencil, and no one would seriously argue that technology should stop with typewriters. However, this does not mean that the cognitive–social transition from typewriters to super word processors must necessarily be as uncoordinated as was that from pencils to typewriters (but cf. Schrodt, 1982).

Indeed, the approach we have just sketched can be elaborated somewhat. It may be too strict a constraint to expect that a single metaphor or a patently obvious metaphor (like the typewriter metaphor) will always be immediately available (or even imaginable) in new areas of technological innovation. So far, we have limited our consideration to monolithic metaphors spontaneously generated by learners as they begin to interact with a system. The possibility is open—and intriguing—that aspects of prior knowledge could be brought to consciousness and assembled into composite metaphors.

The kind of educational technology we are imagining here has simply not been developed. Accordingly, there is little more we can offer at this time. Even if one eschews an active orientation toward the construction of learning metaphors, it seems crucial to guide learners at least to the extent that they should be warned about the limits to which a metaphor can be employed. The implicit typewriter metaphor we identified in our learner's conceptualizations often caused problems for them when it was inappropriate: such cases should be ferreted out and identified for new users.

Exploratory Environments

If the picture of learning we have given here is anything close to correct, it suggests that an optimal learning environment for new users will differ from that provided by current systems. Indeed, the disparity between what exists currently and what would seem to constitute a reasonable learning environment is so great that we are only able to capture the grossest properties of what we imagine would be an optimal learning environment. By definition we will refer to this as an "exploratory environment"; our reason for choosing this term will become clearer presently.

One property of an exploratory environment is that it provides learners with *an appropriate orientation toward the task:* learners approach the task of learning to use a word processor with highly discrepant expectations about what it will be like and how difficult it will be. They—and their bosses—may understand from popular misconception that word processors are easy to learn to use and that using them will result in immediate and substantial productivity increases. This might even be true if the means of instruction is personal tutoring. However, if the means of instruction is contemporary self-study, this is far from the truth—as our studies demonstrate. Realistic expectations about the difficulty of learning to use word processing equipment would at the very least discourage attributions of self-blame on the part of learners. And this, in turn, would at the very least make learning to use these systems more pleasant.

Another aspect of realistic task orientation involves learner responsibility and control. Manuals and system interfaces implicitly attempt to place the learner in a passive role, but as we have seen learners take charge regardless. Nevertheless, the conflict between a manual/system ensemble that directs rote exercises on, say, cursor movement, and an active learner who decides in spite of this to explore techniques of typo correction is problematic: the learner will "fail" in some sense no matter what the outcome. An exploratory environment establishes and reinforces a role of responsibility and control for the learner *via* the system interface and training materials. (Some of our current research work is addressed to this and we return to it briefly in the final section of this study.)

A second property of an exploratory environment is *system simplicity:* learner problems will tangle to the extent that systems afford problem tangling. Systems should, therefore, be designed to minimize potential for tangling. However, this does not at al mean that "function" must be traded for simplicity vis-à-vis the user interface. Earlier we suggested a "staged" sequence for user interfaces. The initial user interface, on this proposal, would indeed trade function for simplicity, but it would do this by shielding the new user from potentially tangling function. In any case, new users are not the ones who complain about limitations of function. Successive interfaces would increase the amount of revealed function—and therefore the potential for problem tangling—but staged in a way that would minimize the risks to learning at any given point in the process.

An alternative approach to achieving this sort of simplicity has been called "progressive disclosure" (Smith, Irby, Kimball, Verplank, & Harslem, 1982). An interface organized this way stages each function separately and places the staging sequence under the user's control. This is a more flexible approach than that of staging the entire interface as a whole and *shielding* the user from control, even awareness, of this. On the other hand, progressive disclosure, although it quite literally makes the learner "ask" for trouble, does allow such requests to be fulfilled.

Simplicitly connects directly to several other properties of exploratory environments, in particular to *clarity* and *safety*. Systems are often unclear in the sense that they do not inform the user of what state they are currently in, what they are currently doing, or even what they must have just done. In one system we studied, even the command to sign-off occasioned no confirmation message at all: the system merely put up the sign-on menu (presumably for the next user). New users do not recognize that "hello" *means* "good-bye" in systems. They frequently signed on again and then asked us how they were ever going to get off of the system!

In the same system, the command to print a document elicited a confirmation message that was routed to a "message reservoir." Here the message waited for the user to request it to be displayed. The trouble is that a user might not realize what has happened and/or might not know how to display a message. This could result in the message being displayed at a time much later than that at which it was sent. Thus when one least expected it, one could learn that file such-and-such had been printed. These problems of the amount and the timing of feedback are rather easily dealt with, at least in principle: why not have more feedback and present it in a more timely fashion? (The relevant design trade off here is that the presentation of feedback might interrupt a succeeding task and thereby distract the user—but this rationale doesn't make printing any less inscrutable for the learner.)

But a further, and far less trivial, feedback problem is that the content of messages is not typically "problem-oriented." Learners do not want to read messages like "parameter omitted or not valid." They want to find out what they did wrong *in that case* with regard to their own current goals—and they would like to know what to try next to achieve those goals. This kind of situation is devastating for current Help systems. We have many times seen subjects examine a Help screen which, in fact, contained the information they needed, but which presented that information in a problem-neutral manner. Since the learners were construing their needs in a problem-oriented "vocabulary," they simply failed to see the relevance of the Help.

Addressing this problem at all would require a task-analysis-driven approach to interface design that regards *learning* as a key user task. Addressing it comprehensively would require building in some sort of "intelligence" to the system interface. Nevertheless, an exploratory environment is "clear" in the sense that it provides abundant, timely, and problem-oriented feedback to the learner.

Safety is the capacity of the system to protect the learner from demoralizing penalty. The issue arises because, as we have seen, the learning strategies people adopt lead them into trouble with current word processors. A system that was safe to learn by doing would be one whose function was organized modularly with respect to real tasks. An error committed while correcting a typo in a menu field would not tangle with reset to alter the applicability of procedures for correcting the typo (recall Table 1). Training materials also should be organized modularly with respect to real tasks: learning by doing does not mean following out a cursor movement exercise by rote; it means typing in a letter (and "incidentally" learning to move the cursor about).

In such a system, recovering from an error might be more than a hopeless jump into an endless morass of problem tangles. Indeed, we view error recovery as a paradigmatic learning situation: the learner is highly task-oriented, primed to take action, and engaged in the system's responses. In current unsafe systems, all this is for naught since the learner is quite likely to end up lost. In a safe system, all this very constructive motivation might be effectively channeled into learning by discovery. Learners must be safe in taking action.

If the strategy of learning by thinking is to operate safely, the inferences, deductions, and abductions that people are prone to make must be correct. In some cases, this point can be codified simply: operations that have apparent natural inverses (space and backspace, line advance and line return) should indeed *be* inverses; they should mutually undo each other's effects; they should have predictably inverted names (Carroll, 1982a). In many other cases, designing systems that can be learned by being reasoned out is not nearly so straightforward. We simply do not understand human reasoning in these complex task environments well enough to derive from principles appropriate design strategies.

And of course, if the strategy of learning by knowing is to be safely employed then care must be taken to present a system interface which is consistent with prior knowledge: if word processors are thought of as super typewriters then, at least the initial interface should contain either nothing inconsistent with this, or if it does contain inconsistent features these should be well-marked for the new user (Carroll & Thomas, 1982). An exploratory environment is safe with respect to the strategies of learning by doing, thinking, and knowing. System operations and training materials are modular with respect to real tasks; natural inferences, deductions, and abductions entail correct expectations about the system's behavior; and prior knowledge which people regard as relevant *is* relevant.

These properties straightforwardly suggest the hypothesis that learning, as we have described it here, is facilitated in an exploratory environment. However, in closing this section, we want to observe that this facilitation may not merely be "cognitive." Exploratory environments may enhance task oriented motivation as well. Computer games manifest many of the properties of exploratory environments, and have been argued to increase task-oriented motivation (Malone, 1980, 1981). The argument is murky, however, since the games differ from word processors in a great many irrelevant ways as well.

The game of Adventure may be an important case with respect to this both because it is so widely popular and because its "maze learning" problem scenario is analogous in many ways to learning the conceptual maze codified in a structure of interconnected menus in a word processor (Carroll, 1982b). Indeed, the similarities between Adventure and current word processing systems can be pushed quite far—which raises the question of why the game is so attractive to people. Part of the answer might be that the types of properties we have clustered into the concept of an exploratory environment—properties like feedback, safety, modularity, learner control—nurture a motivational orientation in game learners that is precluded for learners of current word processors.

Clearly, these are empirical questions that can be addressed only by developing systems that comprise exploratory environments and by studying how people manage to learn to use them. This is the direction of our own current work.

Active Learning

In the early days of cognitive psychology, it was often necessary to rail against "stimulus–response" analyses of learning that went like this: items in the world come around and get paired with responses (or dispositions to respond) in the organism (generally because the pairing led to a good effect of some sort). This is a somewhat barren analysis of human learning to be sure, and cognitive psychology succeeded in doing away with it. The solution offered by cognitive psychology then, and now, was to turn the question of "learning" into the question of "memory." This was probably a good idea; after all, successful learning ought to lead to memory and stimulus–response psychology had nothing to say about how even simple aspects of memory organization could be explained.

In turning the question around like this, however, cognitive psychology overlooked another striking problem with the stimulus–response analysis of learning: learning—in that view—is a mechanically *passive* matter of pairing and then incrementing the strength of pairing. The active initiatives of the learner do not figure in at all. In turning the

learning question into the memory question, cognitive psychology effectively finessed this problem: memory structures, after all, *ought* to be
stable, relatively static, and so on. Reducing the analysis of memory to a
set (albeit an increasingly large set) of mechanisms was viewed as an
entirely proper way to scientifically approach the matter (but cf. Jenkins,
1974).

In sum, *both* stimulus–response psychology and cognitive psychology—in their approaches to the problems of analyzing learning and
memory, respectively—share a basic orientation: the processes involved
are "passive" with respect to the initiative of the learner–memorizer.
They in fact share one other basic orientation. Neither school of psychology focused its principal empirical concern on the analysis of full-scale
problems of human behavior and experience. This is as true of the
classic Sperling and Sternberg brand of memory research as it is of the
classic Skinnerian studies of pigeon pecking. While it may be true that in
highly constrained situations or in infrahuman species active self-initia-
tive is muted, this may not bear at all on the nature of real and complex
human learning.

The treatment of learning in both stimulus–response psychology and
subsequently in cognitive psychology share a picture of learning that
approaches it as passive with respect to the initiative of the learner, and a
methodological orientation that focuses analysis on highly constrained
laboratory situations. Our examination of a more full-scale learning situation questions these premises (as has other recent work in cognitive
science; Norman, 1981). Learning to use a word processor provides a
study of real complex human learning that is fundamentally "active,"
driven by the initiatives of the learner—which are, in turn, based on
extensive domain-specific knowledge and skill, and on reasoning processes which are systematic and yet highly creative.

Where do we go from here? Our examinations of current word processing systems have led us to a conception of learning, and in particular
of learning to use a word processor. In the prior two sections we have
summarized and projected in very general terms what we think we have
learned. However, recognizing that current word processing systems
might be improved from the learner's perspective by providing a conceptual framework and ensuring an exploratory environment is far
short of *understanding what learning would be like in such a situation.* If the
problems that learners experienced were less debilitating, we might be
able to analyze learning at a finer grain and expose cognitive mechanisms that are simply lost in the confusion when people try to learn
current systems. At the least, we could contrast learning in the two
situations.

To address this, we are attempting to codify our suggestions for training and system design into concrete experimental prototypes. A rude approximation of this involves merely altering the training materials learners receive. The system they interact with is one we have studied previously, and one which has very little in common with exploratory environments. Nevertheless, by providing modular, problem-oriented training and training materials which make explicit the basic elements of procedures (e.g., what the goal is, what actions are required, including prerequisites, and what are the outcomes), we have been able to reduce training time without impairing learning performance (as measured by our transfer task).

These are preliminary results (our analysis is in progress) but they are encouraging. The next step for us is to exert direct control over the system interface itself. Only in this way can we seriously assess what learning may be like in an exploratory environment.

The nature of active learning remains for the most part an open question in current cognitive science. Since theory in this area is not very developed or abstract, the most promising strategy seems to be one of representatively examining a variety of accessible and realistic learning domains. Each domain must of course be probed as deeply as possible on its own terms, since we have no way of knowing antecedently which generalizations we identify will turn out to be domain-independent and which will turn out to be properties specific to particular domains or sets of domains. The current study can be placed into this programmatic framework: its chief results are its indications of what needs to be done next.

ACKNOWLEDGMENTS

We are greatly indebted to Clayton Lewis who was a full collaborator in this research and an originator of many of the ideas presented here. He was unable to participate in writing this chapter. We are also grateful to Karen Greer who helped to transcribe from tape-recordings the material presented in tables. Some of this material appeared in a slightly different form in Carroll and Mack (1983).

REFERENCES

Bott, R. *A study of complex learning: theory and methodology.* CHIP Report 82, University of California at San Diego, La Jolla, CA, 1979.

Carroll, J. Learning, using, and designing command paradigms. *Human Learning: Journal of Practical Research and Application. 1*, 31–62, 1982. (a).

Carroll, J. M. The adventure of getting to know a computer. *Computer, 15(11)*, 49–58, 1982. (b)

Carroll, J. & Lasher, M. Getting to know a small computer. Manuscript, IBM Watson Research Center, 1981.

Carroll, J. M., & Mack, R. Actively learning to use a word processor. In W. Cooper (Ed.) *Cognitive aspects of skilled typewriting*. New York: Springer-Verlag, 1983. Pp. 259–281.

Carroll, J., & Thomas, J. Metaphor and the cognitive representation of computing systems. *IEEE Transactions on Systems, Man and Cybernetics, 12*, 107–116, 1982.

Ericsson, K., & Simon, H. Verbal reports as data. *Psychological Review, 87*, 215–251, 1980.

Halasz, F., & Moran, T. Analogy considered harmful. *Proceedings of the Conference on Human Factors in Computer Systems*. National Bureau of Standards, Gaithersburg, MD, 1982.

Hayes-Roth, B., & Hayes-Roth, F. A cognitive model of planning. *Cognitive Science, 3*, 275–310. 1979.

Ingalls, D. Design principles behind Smalltalk. *Byte*, 286–298, August 1981.

Jenkins, J. J. Remember that old theory of memory? Well, forget it! *American Psychologist, 29*, 785–795, 1974.

Kinneavy, J. *A theory of discourse*. Englewood Cliffs, NJ: Prentice-Hall, 1971.

Lewis, C. *Using thinking aloud protocols to study the "cognitive interface"*. Research Report RC 9265, IBM Thomas J. Watson Research Center, Yorktown Heights, NY, 1982.

Lewis, C. & Mack, R. The role of abduction in learning to use text-processing systems. Paper presented at the annual meeting of the American Educational Research Association, New York, March 19–24, 1982.

Mack, R., Lewis, C., & Carroll, J. Learning to use office systems: Problems and prospects. *ACM Transactions on Office Information Systems, 1*, 10–30, July 1983.

Malone, T. *What makes things fun to learn? A study of intrinsically motivating computer games*. Cognitive and Instructional Sciences Series CIS-7 (SSL-80-11). Xerox Palo Alto Research Center, August, 1980.

Malone, T. What makes computer games fun? *Byte*, 258–277, December 1981.

Nisbett, R., & Wilson, T. Telling more than we can know: Verbal reports on mental processes. *Psychological Review, 84*, 231–259, 1977.

Nisbett, R., & Ross, L. *Human inference: Strategies and shortcomings of social judgment*. Englewood Cliffs, NJ: Prentice-Hall, 1980.

Norman, D. A. (Ed.). *Perspectives on cognitive science*. Norwood, NJ; Ablex, 1981.

Peirce, C. S. The logic of drawing history from ancient documents. In A. Burks (Ed.), *Collected papers of Charles Sanders Peirce*. Cambridge, MA; Harvard University Press, 1958.

Rumelhart, D., & Ortony, A. The representation of knowledge in memory. In R. C. Anderson, R. Spiro, & W. Montague (Eds.), *Schooling and the acquisition of knowledge*. Hillsdale, NJ; Erlbaum, 1977.

Schrodt, P. The generic word processor: A word-processing system for all your needs. *Byte*, 32–36, April 1982.

Smith, D., Irby, C., Kimball, R., Verplank, B., & Harslem, E. Designing the STAR user interface. *Byte*, 242–282, April 1982.

Zloof, M. Query-by-example: A data base language. *IBM Systems Journal, 16*, 324–343, 1977.

3

Formal Grammar as a Tool for Analyzing Ease of Use: Some Fundamental Concepts

PHYLLIS REISNER

> Human factors is still an
> experimental science, not a predictive
> one.
>
> L. Branscomb
> *IBM chief scientist (1981)*

This chapter presents a method for describing the design of a man machine interface before the interface is available for behavioral testing. The method starts by describing the user's "action language" with a formal grammar. It then gives an explicit procedure for making predictions from the grammar. Two key concepts are introduced, the notion of *cognitive terminal symbol* and the notion of *prediction assumptions*. Cognitive terminal symbols are used to make user's "thinking actions" explicit in the grammar. Prediction assumptions are used to adapt predictions to the user population, thus making different predictions, for the same interface, when the user populations differ. A detailed example of the technique is given in this chapter.

To find out whether the human interface to a computer system is easy to use, human factors practitioners run experiments which measure human performance. These experiments frequently require that a working version of the interface already exist to give the experimenter something to test. The testing will then come very late in the development cycle. If the interface is not easy to use, either the system is redesigned or the experimental results are ignored. Redesign is costly for the developer. Ignoring poor human factors is costly in the market place.

This situation is both inefficient and intellectually unsatisfying. It is akin to that of a bridge builder who sends people across a bridge to determine whether it will support its intended weight. A better test procedure would be to analyze the bridge design mathematically. Analogously, we would like to be able to analyze the design of a man–machine interface, on paper, before a working model is available for behavioral testing. Such analysis should be used to locate flaws in the design of an interface that would create problems for users. It should also be used to compare alternative designs for ease of use.

We wish to develop human factors into a predictive science as well as an experimental one. Therefore, we need tools for examining the design of an interface before it is available for behavioral testing. Such tools should be *analytic,* permitting us to describe the design of an interface on paper, then manipulate the description to assess its ease of use. Earlier work (Reisner, 1981) showed that formal grammar could be used as such an analytic tool to describe a user "action language" (the sequence of keypresses, cursor motions, etc., used as input to a system). It then showed how this formal description could be examined to compare two designs for ease of use and to identify possible user errors. This early work, while relatively concrete, requires further development before it can be used as a design tool.

In the current chapter, therefore, we describe an explicit, and general, prediction process. This process eventually uses simple computation to make predictions about ease of use. Several new concepts are used in this process:

1. We show how the action language idea can be extended to include cognition (knowing). What a person has to know in order to use the interface to a computer system is an important factor in ease of use. A formal description intended for analyzing ease of use should permit us to represent cognition in a useful way.
2. We show how interface designs can be compared, based on "sentences" obtained from formal grammars to describe specific action sequences. This will help to make the prediction process explicit.
3. We show how a grammatical description of an interface can be related to predictions of time or errors. This, too, will help to make the prediction process explicit.
4. We introduce the concept of *prediction assumptions.* This concept attempts to capture, explicitly, the knowledge of a human factors expert about human abilities. It does so in a way which can be manipulated by simple mathematical computation. The concept, furthermore, permits us to tailor predictions about ease of use to specific user populations. It is a truism that what is easy for one kind of user may be difficult for another. The prediction assumptions permit us to make different predictions about ease of use, for the same interface, when the user populations are different.

In this chapter, we first review some related work. We then introduce the new concepts and show how they can be combined into an explicit prediction process. Following that, we give detailed examples of predictions using the process. The main example compares different ways of identifying a line in a text editor (using a "locate" command or "scrolling" with several options). Finally, we discuss a number of implications of the work. This chapter discusses work that is still in progress.

BACKGROUND

There are three research threads that come together in this work: the notion of user actions as a "language," the use of formal grammar to describe the structure of languages in linguistics and in computer science, and the use of behavioral tests of models of user–computer communication. The first research thread, that user input at a terminal can be viewed as a language is, implicitly, embodied in the jargon of computer science. (A man–machine "conversation" presupposes a language.) The notion has been clearly expressed in (Foley & Wallace, 1974; Foley, 1980). By viewing such user input as a language, we presuppose that underlying the strings of input actions there is a *structure,* that this structure can be described explicitly by "syntactic rules" and that this structure is important to ease of use.

The second thread is that of formal grammar. Linguists, of the "generative grammar" school, describe natural languages with a set of "production rules." These rules are assumed to be those that a native speaker of the language follows implicitly in creating and in understanding languages (Chomsky, 1964). Computer scientists also use formal grammars (e.g., BNF). They use the grammars to describe programming languages to translators (e.g., compilers) or to compiler generators (Lewis, Rosenkranz, & Stearns, 1976). One very important aspect of this formalism is that it is relatively succinct. It can be used to describe all possible sentences in a language—a possibly infinite set—with a finite set of rules.

The third thread, models of man–machine interaction, is just becoming visible. One example is the keystroke-level model of Card and Moran (1980). This model describes the time it will take expert users to use a text editor. After fitting parameters to the model, Card and Moran test it by comparison with the behavior of expert users. Embley, Lan, Leinbaugh, and Nagy (1978) have a similar model based on keystroke counts.

Other work is related to some of the above research threads. Moran (1981) presents a formalism he has devised, CLG (Command Language Grammar) to describe all levels of a system from the conceptual to the physical device level. Other descriptions of the user interface by some formal method are: Parnas (1969); Feyock (1977); Embley (1978); Ledgard and Singer (1978); Shaw (1980); Wasserman and Stinson (1981); Bleser and Foley (1982); Jacob (1982); Schneiderman (1982);

Kieras and Polson (1982). An early suggestion that BNF description might be used to predict *psychological* difficulty of user languages (query languages) can be found in Reisner (1977).

In Reisner (1981), these three threads were woven together. User input actions at a terminal were viewed as an "action language." Using a small experimental color graphics system (ROBART) as an example, a formal grammar for the action language was written, using the well-known BNF or production rule notation, but applying it to user actions, rather than to programming language or to natural language. Then the formal description was examined to identify design flaws and to compare two versions of ROBART for ease of use. The predictions made were quite explicit. For example, one predictions was that users would make a specific error (press a button labeled GO) at a specific point in the process of trying to draw etch-a-sketch type lines. Another prediction was that users would find it harder to remember how to choose a shape (e.g., line, circle, or rectangle) in the first version of ROBART than in the second. The predictions were substantiated by experiments. Readers unfamiliar with formal grammar can find a very brief description, including "how to read the notation," in Reisner (1981).

FOUNDATIONS

In this section we discuss the four concepts which improve the descriptive and predictive power of the formal grammar. In the next section, these concepts will be integrated into an overall prediction process. We discuss: (*a*) introducing cognition into the formal description; (*b*) comparing two designs; (*c*) relating the grammatical model to measurable quantities; and (*d*) the notion of prediction assumptions. The most important points are the introduction of cognition and the notion of prediction assumptions. The other two points are needed simply to make the prediction process explicit.

Introducing Cognition into the Formal Description

Motivation

There are at least two requirements for a formal notation that is to be used to predict ease of use. The first requirement is that the description include all information that we know to be central to ease of use. Cognition, in particular, because it plays such an important part in determining whether or not a system is easy to use, *must* be represented explicitly in the grammar. Furthermore, it should be represented in a way that can be manipulated and differentiated from other factors affecting ease of use.

Second, the terminal symbols (words) of the grammar should be operationally defined (or definable): that is, an analyst should have a specific procedure, or set of operations, to follow in identifying the terminal symbols. The result of following these rules should always give the same result. Thus, two analysts, following the rules, should develop grammars with the same terminal symbols.

While the importance of cognition was recognized in the earlier work, it was not represented in the ROBART grammars in a way that could be operationally defined or independently manipulated. In the ROBART work, terminal symbols were called "cognitive terminal symbols" and the grammar was called a "cognitive grammar." These terms were explained, but not operationally defined. A ROBART "cognitive terminal symbol" represented "actions the user has to learn and remember." Some examples were:

> PUT LINE SWITCH UP
> POSITION CURSOR AT CIRCUMFERENCE

In the ROBART work, the notion of cognitive terminal symbol was only defined by example. Since it was not operationally defined, two analysts could easily have developed grammars for ROBART with different terminal symbols. Furthermore, since there was only one kind of terminal symbol, the effect of cognition on ease of use could not be distinguished in the grammar from the effect of physical input actions. Thus their relative importance could not be represented in the prediction method.

Approach

To meet these objectives, we distinguish *two* different kinds of symbols: physical input symbols and cognitive symbols. Specifically, we focus here on the terminal symbols, although the distinction between physical input symbol and cognitive symbol also applies to nonterminals. The physical input symbols represent user actions that (a) are observable, and (b) serve as input to the system, e.g., pressing a key, turning a knob, pointing with a light pen, moving a switch. These are clearly operationally definable.

The second class of terminal symbols represents cognition. However, representing "actions the user has to learn and remember" (*what* the user has to know—the approach taken in the ROBART grammars) combines actions with knowledge. It thus does not lead to a coherent model. A more productive way of handling cognition is, in fact, implicit in the notion of "user action language." Since we are describing user actions in these grammars, the class of symbols representing cognition should represent user *cognitive actions*. Cognitive actions are mental "behaviors" such as performing a mental calculation, remembering the syntax for a command, or deciding between alternative forms of a command. Where the ROBART grammars did not distinguish between mental and physical actions, both kinds of actions can now be explicitly represented. Thus, the ROBART terminal symbol PUT LINE SWITCH UP would be replaced by two symbols, one representing cognition, the other, physical input: <REMEMBER HOW TO SELECT A LINE> and PUT LINE SWITCH UP. We will enclose cognitive symbols in brackets, as shown. In some cases, instead of merely adding a cognitive symbol, as in the above example, we will both add symbols and change the existing ones. For example, the ROBART symbol POSITION CURSOR AT CIRCUMFERENCE does not represent either a cognitive action or a physical one. It represents "what" the user has to know (that circle size is determined by choosing a point on the circumference). It also represents the *result on the screen* of a physical input action (the cursor moves on the screen as a result of the physical input action of moving a joystick in the ROBART system). Consequently, the symbol POSITION CURSOR AT CIRCUMFERENCE could be replaced by the cognitive actions <REMEMBER HOW TO IDENTIFY CIRCLE SIZE> and <CHOOSE CIRCUMFERENCE POINT>, and by the physical input action, MOVE JOYSTICK.

A simple example of part of a grammar including cognitive actions is shown in Figure 1. This example represents the actions required to delete a block of consecutive lines in a text editor with a command, "Dn" (delete *n* lines). Terminal symbols ("words") are represented in capital letters; nonterminals (the equivalent of "parts of speech") in small letters. Cognitive symbols are represented in brackets. In addition, the anomalous class of information seeking actions, such as looking up information in a book, is also represented in brackets. These actions are clearly physically observable, but do not represent input to the computer system. Discussion of the conventions for nonterminals is deferred to a later paper.

Defining two kinds of symbols, physical input symbols and cognitive symbols, brings us closer to meeting the objectives stated above. Cognition has been represented, explicitly, in the notation. It can be independently manipulated. Furthermore, we can compare the effect on ease of use of cognitive actions with input actions. We can also compare the effect of different kinds of cognitive actions with each other. How we do this will be shown later.

In addition, the terminal symbols representing physical input actions are operationally definable. We suspect strongly that the terminal symbols representing cognitive actions can also be operationally defined, but

```
employ Dn                        ->  <retrieve info. on Dn syntax>
                                     + use Dn
<retrieve info. on Dn syntax>    ->  <retrieve from human memory>
                                     |<retrieve from external source>
<retrieve from human memory>     ->  <RETRIEVE FROM LONG TERM MEMORY>
                                     | <RETRIEVE FROM WORKING MEMORY>
                                     | <USE MUSCLE MEMORY>
retrieve from external source    ->  RETRIEVE FROM BOOK | ASK  SOMEONE
                                     | EXPERIMENT | USE ON-LINE HELP
use Dn                           ->  identify first line
                                     + enter Dn command
                                     + PRESS ENTER
identify first line              ->  ...
enter Dn command                 ->  TYPE D + type n
       ...
```

FIGURE 1. **Part of a grammar for a delete command in a text editor, showing cognitive actions. Notational conventions are described in the text.**

this is yet to be proved. We suspect that there is a small number of such actions, which can be defined by listing them (definition "by extension"). We also suspect that the presence of such actions can be inferred from the physical actions which follow them. For example, in the command "Dn" mentioned above, the number n must be obtained by some kind of mental operation such as counting or calculating.

Comparing Two Designs

Given two different designs for the same function, it is necessary to represent each one explicitly in order to compare them. To do this, we need to distinguish the notion of a grammar from the notion of a sentence.

In linguistics, a grammar represents an entire language. Specific sentences are derived from the grammar by selection of appropriate rules for that sentence. By analogy, a grammar of an action language represents the entire language. The designs that we want to compare are represented by different sentences derived from that language. Thus, to compare the difficulty of two ways of deleting lines in a text editor, the grammar would have rules showing that there were two ways of deleting lines. Two sentences are then derived from the grammar, one for each of the delete methods. Predictions about ease of use are then made, using the methodology to be described later, based on these two sentences.

This simple distinction permits us to make predictions that go beyond comparison of specific functions. We are clearly able to compare combinations of functions, e.g., a "delete" function followed by an "insert" function in a text editor. In addition, there is another benefit to this distinction. We can also compare different populations of users. Since the rules of the language can represent different possible user actions, different choices can be made from this grammar, depending on who the user is. For example, one kind of user action is to determine the syntax for the delete command mentioned above. Different users might determine this syntax differently. A novice user, for example, might use an on-line HELP system, if it exists, or look up information in a book. An expert, on the other hand, might already know the syntax. His "cognitive action" would be to remember the information (retrieve it from his long-term memory). These different ways of determining the syntax would be embodied in different rules in the grammar. Different sentences would then be derived from the grammar, one for the novice, and one for the expert. The same approach can be used to describe different conditions under which a system might be used, e.g., the different stages a user passes through in learning a system or how recently he has used it. In these two cases, as in the case of the novice and expert user, the method for determining the syntax might differ, and different choices would then be made from the grammar to represent these differences. The notion of "prediction assumptions," which we will introduce later, is another way of adapting the predictions to the user population.

Relating a Grammatical Model to Measurable Quantities

Motivation

The next problem to be dealt with is how to relate a grammatical model to time or errors. The elements of a grammatical model are, for example, the equivalents of words, parts of speech, rules of syntax, or sentences. Ease of use, on the other hand, is measured in terms of time or user errors. Predictions about ease of use should be made in terms of these quantities. We thus need to establish a method for linking the grammatical model to the quantities to be measured.

In the ROBART work this link was only implicit, not explicit. For example, a prediction was made that selecting a shape (e.g., circle) would be "easier" in one system than the other, and then the number of user errors was measured. The step between "easier" and "fewer errors" is an obvious, but still an inferential one.

Approach

Again, the solution to this problem is implicit in the fact that we are describing a user action language. The terminal symbols in the grammar represent user *actions*. The actions are either physical actions or cogni-

$$S \rightarrow Word_A + Word_B + Word_C + Word_D$$
$$| \qquad | \qquad | \qquad |$$
$$t_S = t_A + t_B + t_C + t_D$$

FIGURE 2. Associating a "sentence" in a user action language with an equation whose variables are times to perform the actions.

tive ones. Any action takes some amount of time and has associated with it some probability of error. Therefore, we simply associate each terminal symbol with a variable representing the time to perform the action or the probability of error associated with it. We can do this in the grammar itself or in sentences derived from the grammar. To do this in the grammar, we simply add a set of rules which rewrite the previous terminal symbols as variables. Thus, if the grammar previously had a terminal symbol, MOVE JOYSTICK, the new grammar would have a rule,

$$\text{move joystick} \rightarrow t_{\text{MOVE JOYSTICK}}$$

If we prefer to work with sentences instead of with the grammar, we would simply associate each word in the sentence with the corresponding time. Thus if we had a sentence consisting of words, A, B, C, D, we would associate with it an equation with terms t_A, etc., as shown in Figure 2. Errors can be handled in an analogous way. Associating the grammatical model with time or errors is another step toward the clean, explicit prediction process we want to develop.

Prediction Assumptions

Motivation

When analysts make a judgment that one system will be easier to use than another, this judgment is usually based on some knowledge or underlying assumptions about human performance. Such knowledge is not necessarily consciously formulated or explicit. It can range from an intuitive sense of how difficult it is to perform some particular mental or physical actions to data or guidelines in generally accepted reference manuals. For equipment design in general, examples of such manuals are McCormick (1970), Van Cott and Kincade (1972), and Woodson and Conover (1973). The skill of an analyst may depend on how well he or she can use this knowledge. Since we want an explicit prediction method, however, such "expert knowledge" needs to be stated explicitly. We attempt to capture such knowledge in the notion of "prediction assumptions."

There is another motivation for the prediction assumptions. Whether or not a system is easy to use depends on the intended user. A system that a novice user will find easy to use may be frustratingly slow for an experienced one. A system which is easy for a skilled typist may be difficult for the hunt-and-peck typist or the person who balks at using a typewriter at all. Predictions about ease of use must be able to take such differences in user abilities into account. A prediction method must be able to make different predictions, for the same system, when the user population differs. The prediction assumptions permit us to do so.

There are several requirements that these prediction assumptions should meet. We are aiming toward a rigorous, perhaps even automatable, prediction process. Therefore, we want to express these assumptions about user abilities in a notation which can be manipulated by some mechanizable process. And, since we are going to want to make predictions in terms of measurable quantities, we would like to express this knowledge in terms of time or errors.

In the ROBART work, the assumptions underlying the predictions were not stated. In fact, the assumptions were so intuitive that we were not even aware that they existed. There were two criteria that were used to make predictions: the length of the terminal strings (number of actions in a sequence) and the number of "extra" rules in the grammar. The second criterion, number of extra rules in the grammar, was taken as an indicator of inconsistency in the language. Such inconsistency can be a cause of user error. The concept will be illustrated by an example from English.

In English, there are a number of ways of creating the past tense of a verb. For the so-called "weak verbs," the letters "-ed" are added to the present tense. Thus, the past tense of the verb "walk" is "walked." This is represented by one rule in a grammar of English. But for the so-called "strong verbs," a major change in the word stem occurs. Thus, the past tense of the word "go" is "went." This is represented by a second rule in English. For a new learner of English, the language is therefore inconsistent. There are two rules for creating the past tense instead of one. This inconsistency in the language does indeed create problems for the new user. In this case the "new user," about two years old, will spontaneously say "yesterday I go-ed to the park," using the wrong rule (unconsciously, of course) for creating the past tense. Analogously, such inconsistency in an action language for computer users can also be represented by extra rules.

Underneath the two criteria used for the ROBART predictions were underlying, implicit assumptions. (The longer the string, the more time it will take. The more extra rules, the greater the chance of user error.)

The idea of "prediction assumption" generalizes the first of these criteria.

Approach

We proceed as follows. We express the underlying assumptions about ease of use in terms of simple mathematical inequalities or equations. We derive these statements from many sources: common sense, experimental findings in the human factors or the psychological literature, or behavioral experiments suggested by the analysis itself. Some examples of underlying assumptions are (all else considered equal):

1. Time to retrieve information from an external source such as a book will be greater than time to retrieve from human long-term memory (i.e., $t_{BOOK} > t_{LTM}$)
2. Time to type a particular string of n characters will be less than time to type a string of $n+C$ characters, where C is a positive integer (i.e., $t_{\text{type } n} < t_{\text{type } n+C}$).
3. Time to perform the mental action of copying a number one sees in front of him will be less than time to perform the mental action of calculating the number according to a (specific) given formula ($t_{\text{copy } n} < t_{\text{calculate } n}$).

It will sometimes be useful to state that some difference is expected to be relatively large, namely:

4. $t_{\text{calculate } n} >> t_{\text{type } n}$

The above assumptions may not be true in all cases. However, by stating them explicitly, we can determine when they are true, and when not, by simply thinking them through, by discussion, or by behavioral experimentation. The main point is that the assumptions must be made explicit. Only when they are explicit can they be accepted, refuted, or modified. Once they are explicit, we can decide whether they do or do not apply to a particular user population and system. Furthermore, once we begin to make the assumptions explicit, other assumptions become readily apparent. We have thus a much larger basis for making statements about ease of use then the two criteria previously used. In addition, we expect that the prediction assumptions will be fairly general, thus applicable to a wide variety of interfaces.

Some prediction assumptions deal with cognitive actions, some with physical ones, and some with a mixture. Furthermore, different kinds of cognitive actions are differentiated from each other. We are thus clearly in a position to manipulate these quantities independently.

PREDICTION PROCESS

We now have all the ingredients for an explicit prediction process. The process incorporates the concepts discussed above. It starts with a user action language which includes cognition and eventually makes predictions by means of simple mathematical computations. It can be used to make different predictions for different kinds of user populations.

We first describe the process, then give two examples. The first example is a simple comparison of naive and expert users. We use this comparison primarily to illustrate the methodology. The second example is more detailed. It is a comparison of different possible strategies for identifying a line in a text editor.

The process is as follows:

1. Describe the user action language with a formal grammar. Include cognitive actions in the grammar. (Alternatively, include cognitive actions in the sentences derived from the grammar in step 2.)
2. Derive sentences (or strings) from the grammar, for the specific aspects of design to be compared.
3. Convert the sentences to equations, with time or errors as the variables.
4. State the prediction assumptions explicitly, as inequalities or equations.
5. Substitute the prediction assumptions in the equations and solve the resulting equations with simple algebra.

EXAMPLE: COMPARING NAIVE AND EXPERT USERS

It is well known that naive users will take longer than expert users to perform some tasks. However, we would like to know that our formalism is capable of expressing facts that we know to be true. To illustrate the method, we compare naive and expert users, of a Dn command in a text editor. We only give enough of the grammar to illustrate the approach. We abbreviate the terms slightly when the meaning is clear.

Step 1. *Write a minigrammar.* The grammar is shown in Figure 1. Notational conventions were described above ("Introducing Cognition into the Formal Description").

Step 2. *Derive sentences (or strings) from the grammar.* For illustration, we can assume that naive (first-time) users will have to consult an external source such as a book for the syntax, while expert users can be expected to know it (retrieve it from their own memory). It is not necessary to fully derive sentences for this comparison. We can stop with intermediate strings as shown. This is possible because the prediction assumptions which we will use later apply to all relevant classes. For example, they apply to all kinds of retrieval from external sources, not just to retrieval from a book. It is also possible because the physical input actions required to enter the command are the same in both cases we want to compare. If we wished to derive sentences instead of stopping the derivation with intermediate strings, we would make some choice, e.g., BOOK, for the naive user, and another, e.g., RETRIEVE FROM LONG-TERM MEMORY, for the expert. We would also substitute the actual physical input actions for the Dn command.

We proceed as follows. We generate two strings, one for the naive users, another for the experts. The double arrow is used to indicate a derivation.

NAIVE
employ Dn $\Rightarrow$ <retrieve info. on Dn syntax>
                                                + use Dn
                     $\Rightarrow$ <retrieve from external source>
                                                + use Dn

EXPERT
employ Dn $\Rightarrow$ <retrieve info. on Dn syntax>
                                                + use Dn
                     $\Rightarrow$ <retrieve from human memory>
                                                + use Dn

Step 3. *Convert the strings to equations.* We then convert the strings to equations.

$$t_{Dn \text{ (naive)}} = t_{\text{retrieve from external source}} + t_{\text{use } Dn} \tag{1}$$

$$t_{Dn \text{ (expert)}} = t_{\text{retrieve from human memory}} + t_{\text{use } Dn} \tag{2}$$

Step 4. *State the prediction assumptions.* In this case, the relevant prediction assumption can simply be drawn from common sense. We would expect that retrieval of information from an external source would take longer than retrieval from human long-term memory, namely,

$$t_{\text{retrieve from external source}} > t_{\text{retrieve from human memory}} \tag{3}$$

Step 5. *Substitute and solve.* We then convert the inequality (3) to an equation, where C is a positive integer. (We spell out the next steps only for the benefit of the occasional reader whose acquaintance with elementary algebra may be a bit rusty.)

$$t_{\text{retrieve from external source}} = t_{\text{retrieve from human memory}} + C \tag{4}$$

Substituting Eq. (4) in Eq. (1), we get,

$$t_{\text{D}n \text{ (naive)}} = t_{\text{retrieve from human memory}} + C + t_{\text{use D}n} \qquad (5)$$

But we already know from Eq. (2) that,

$$t_{\text{D}n \text{ (expert)}} = t_{\text{retrieve from human memory}} + t_{\text{use D}n} \qquad (2)$$

So it is clear, by comparison of Eqs. (2) and (5), that experts will be faster than naive users. This is not startling. However, it is clear that we can express, *in the notation,* facts that we know to be true.

EXAMPLE: COMPARING USER STRATEGIES

Model with Physical Input Actions Alone

In this more complex example, we illustrate how the methodology described above can be used to compare different strategies for using the Dn command. By "different strategies," we mean different ways a user can go about accomplishing the same goal. We are also interested in seeing whether including cognition in the description makes a difference in the predictions. Consequently, we first illustrate the methodology using a grammar with only physical input symbols. We then include cognition in the description and compare the predictions with and without cognition.

In this example, we compare different user strategies for identifying the first line in a command such as the Dn command. To "identify" a line, we mean move it onto the screen and move the cursor to the first column of the line, the position where the command is to be typed. Among many possible strategies, we compare the following:

1. Locate: The user types a locate command in a command input field (e.g., L300). Line 300 then moves to the top position of the text area on the screen.

2. Scroll exact: The user first types a parameter number into a scroll field on the screen to indicate the amount to be scrolled (moved) when a particular specialized function key is pressed. In the case of the scroll exact strategy, the parameter number is the exact amount to be scrolled. Pressing a function key (SCROLLDOWN) once then moves the target line to the top position of the text area on the screen.

3. Scroll repeat: The user does not enter a parameter number. The default amount is thus the amount scrolled. The user presses the scroll key (SCROLLDOWN) repetitively until the target line is positioned on the screen.

Some further details are necessary.

1. The derivation applies to an IBM 3278 terminal with 43 lines available on the screen. The first two are command lines. The second of these lines contains the command input field on the left and the scroll field on the right. The remaining 41 of the lines are available for text. Lines at the bottom of the screen are reserved for system status and are irrelevant to this discussion.

2. We assume that the cursor is not at the command input area to start. A specialized function key, PFCURSOR, is used to move the cursor to important points on the screen. When the cursor is in the text area, one press of the PFCURSOR key moves the cursor to the command input field. Two presses moves it to the scroll field. Furthermore, there will be cases in which the cursor is at some line in the text portion of the screen, and some action causes that line to move off the screen. In these cases, the cursor moves to the command input field.

3. Once the target line is on the screen, the cursor must be positioned to enter the Dn command. We assume in all cases that the CARRIAGE-RETURN key is used to do so. For the locate and scroll exact strategies, this key must be pressed once since the desired line is now at the top of the text portion of the screen. For the scroll repeat strategy, it must be pressed repetitively when the line is not at the top of the screen.

4. The default scroll value is half the number of available text lines. In the implementation we are using, this value is 20 lines.

5. We will assume for the example that the top position of the text area on the screen contains an arbitrary line of the text, line 5. We do so because having the first line of the text in the first position is not typical.

6. If the cursor is in the command input field, or in the scroll field, it stays there when an entry is made in that field.

Procedure

The procedure is as follows.

Step 1: *Write a grammar.* A grammar is given in Appendix A. It includes only physical input symbols.

Step 2: *Derive sentences from the grammar.* We generate "typical" sentences from the grammar for the strategies to be compared. To do so, we pick a "typical" task: in this case, "delete lines 300 to 800." We then generate sentences for these tasks from the grammar. Later, we examine the results to see whether they depend on the particular details chosen as typical.

The derivation for the locate strategy is shown in Figure 3. To illustrate another method of representing a derivation, we show the derivation tree for the scroll exact strategy in Figure 4. We omit the words

```
use Dn  => identify first line + enter Dn command +  ENTER
        => get first line on screen + move cursor to first line
           + enter Dn command + ENTER
        => use locate strategy + move cursor to first line
           + enter Dn command  + ENTER
        => move cursor to command input field + type locate command
           + ENTER +  move cursor to first line + enter Dn command + ENTER
        =>  PFCURSOR +  type locate command + ENTER
           + move cursor to first line + enter Dn command + ENTER
        => PFCURSOR +  type locate keyword + type line number
           + ENTER + move cursor to first line + enter Dn command + ENTER
        => PFCURSOR + L +  type number + ENTER
           + move cursor to first line +  enter Dn command + ENTER
        => PFCURSOR + L + (3 + 0 + 0) + ENTER
           + move cursor to first line + enter Dn command + ENTER
        => PFCURSOR + L + (3 + 0 + 0) + ENTER
           + move cursor to first line + D
           + type numerical difference + ENTER
        => PFCURSOR + L + (3 + 0 + 0) + ENTER
           + move cursor to first line + D
           + type number + ENTER
        => PFCURSOR + L + (3 + 0 + 0) + ENTER
           + move cursor to first line
           + D + (5 + 0 + 1) + ENTER
        => PFCURSOR + L + (3 + 0 + 0) + ENTER
           + use cursor keys + D + (5 + 0 + 1) + ENTER
        => PFCURSOR + L + (3 + 0 + 0) + ENTER
           + CARRIAGERETURN + D + (5 + 0 + 1) + ENTER
```

FIGURE 3. **Derivation for locate strategy.**

PRESS and TYPE for brevity. Also for brevity, we indicate that the PFCURSOR key is to be pressed twice by "(2x)."

We obtain the derivation for the scroll repeat strategy by either of the above methods. Some further details for the scroll repeat strategy are necessary. To calculate the number of times the scroll key (SCROLL-DOWN) is pressed, we use: (target line − line-in-top-position)/scroll amount, i.e., $(300 − 5)/20 = 14+$. Fourteen keypresses brings line 300 into position on the screen. The first line of the text area on the screen will contain line 285 $[(20 \times 14) + 5]$. Since the old line has moved off the screen, the cursor moves to the command input line. The CARRIAGERETURN key must then be pressed 16 times to move the cursor to the line 300. $[(300 − 285) + 1]$.

The sentences representing the three strategies are given below.

1. Locate Strategy

PFCURSOR + L + 3+0+0 + ENTER + CARRIAGERETURN + D + 5+0+1
+ ENTER

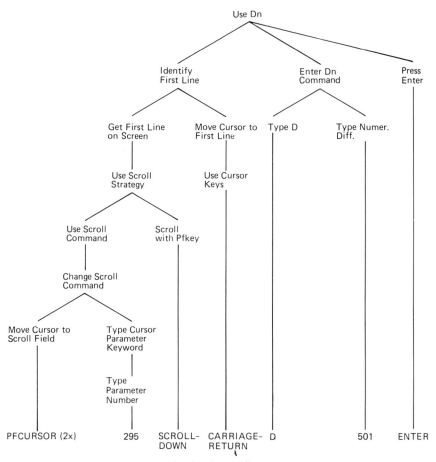

FIGURE 4. **Derivation for scroll exact strategy.**

2. Scroll exact strategy

PFCURSOR(2x) + 2+9+5 + SCROLLDOWN + CARRIAGERETURN + D + 5+0+1
+ ENTER

3. Scroll repeat strategy

NULL + SCROLLDOWN(14x) + CARRIAGERETURN(16x) + D + 5+0+1
+ ENTER

Step 3. *Convert the sentences or strings to equations.* The equations are:
1. Locate strategy

$$t_{\text{loc}} = t_{\text{PFCURSOR}} + t_{\text{L}} + t_3 + t_0 + t_0 + t_{\text{ENTER}}$$
$$+ t_{\text{CARRIAGERETURN}} = t_{\text{D}} + t_5 + t_0 + t_1 + t_{\text{ENTER}}$$

2. Scroll exact strategy

$$t_{\text{scrollexact}} = 2t_{\text{PFCURSOR}} + t_2 + t_9 + t_5 + t_{\text{SCROLLDOWN}}$$
$$+ t_{\text{CARRIAGERETURN}} + t_D + t_5 + t_0 + t_1 + t_{\text{ENTER}}$$

3. Scroll repeat strategy

$$t_{\text{scrollrepeat}} = 14t_{\text{SCROLLDOWN}} + 16t_{\text{CARRIAGERETURN}} + t_D$$
$$+ t_5 + t_0 + t_1 + t_{\text{ENTER}}$$

Step 4. *State the prediction assumptions.* We will assume that typing any key, or pressing any function key, takes the same amount of time as any other. We will also assume that typing a string of length n will take more time than typing a string of length $n + C$, where C is a positive integer, and that typing two strings of length n will take the same amount of time. Thus, where t_{KEY} is the time to type any key, the following is true.

$$t_{\text{locate}} = 12t_{\text{KEY}}$$
$$t_{\text{scrollexact}} = 12t_{\text{KEY}}$$
$$t_{\text{scrollrepeat}} = 35t_{\text{KEY}}$$

Step 5. *Substitute and solve.* The above assumptions are the equivalent of simply counting string length. Longer strings are assumed to take more time. Substituting in the equations is unnecessary, since the result is obvious by inspection.

Predictions from Physical Input Actions Alone

How do these three strategies compare in ease of use? If we count string length, the lengths are 12, 12, and 35, respectively. Therefore, a model based only on physical actions would predict:

P1. There will be no difference between the locate and the scroll exact strategies.

P2. The scroll repeat strategy will take more time than the either the locate or the scroll exact strategies.

There will be minor cases where these predictions will not be true. The scroll exact strategy will have one action less than the locate strategy when the difference between the target line and the top line (here, 295) has fewer digits than the target line (here, 300). Also, when the target line number is small and scrolling brings it near the top (e.g., 47), thus requiring very few repetitions of the SCROLLDOWN and CAR-RIAGERETURN keys, the scroll repeat strategy will be faster than the scroll exact strategy. These are anomalous cases, however, the should not make a difference in the overall prediction.

Model with Physical *and* Cognitive Input Actions

The simple physical action model just described does not capture differences between the strategies which we intuitively believe must exist. The differences lie in the mental operations people must perform. The above approach predicted no difference between the locate and the scroll exact strategies since each requires the same number of keypresses. We want to see whether addition of cognition changes these predictions. We therefore expand the sentences to include some "cognitive" actions.

Given the task, "delete lines 300 to 800," a person using the locate strategy only has to copy the number 300 after the symbol "L" in order to locate line 300. So the cognitive action is "copy number." To enter the scroll parameter (295) in the scroll exact strategy, the cognitive action is "calculate number" (target line − line-in-top-position). There is also a cognitive action to determine the number of lines to be deleted (501). This is also a calculate action, but different from the first ((last line − first line) + 1). We will call the two calculations calc1 and calc2, respectively. The new sentences, rewritten to substitute a pair of symbols (cognitive symbol + physical symbol) for the relevant physical symbols, are written below.

Procedure

The expanded sentences are:

Locate strategy

> PFCURSOR + L + <COPY(300)> + 3+0+0 + ENTER +
> CARRIAGERETURN + D + <CALC2(501)> + 5+0+1 + ENTER

Scroll exact strategy

> PFCURSOR(2x)+ <CALC1(295)> + 2+9+5 + SCROLLDOWN +
> CARRIAGERETURN + D + <CALC2(501)> + 5+0+1 + ENTER

We convert the sentences to equations.

$$
\begin{aligned}
t_{\text{locate}} = {} & t_{\text{PFCURSOR}} + t_{\text{L}} + t_{\text{COPY(300)}} + t_3 + t_0 + t_0 \\
& + t_{\text{ENTER}} + t_{\text{CARRIAGERETURN}} + t_{\text{D}} + t_{\text{CALC2(501)}} \\
& + t_5 + t_0 + t_1 + t_{\text{ENTER}}
\end{aligned}
\tag{1}
$$

$$
\begin{aligned}
t_{\text{scrollexact}} = {} & 2t_{\text{PFCURSOR}} + t_{\text{CALC1(295)}} + t_2 + t_9 + t_5 \\
& + t_{\text{SCROLLDOWN}} + t_{\text{CARRIAGERETURN}} + t_{\text{D}} \\
& + t_{\text{CALC2(501)}} + t_5 + t_0 + t_1 + t_{\text{ENTER}}
\end{aligned}
\tag{2}
$$

The predictions assumptions follow. For this analysis, again, we will assume that time to press any key is the same as time to press any other key. For finer grained analyses, we could, of course, make different assumptions. We also want to use the cognitive terminal symbols we introduced earlier. To do so, we state an assumption that calculating a number will be slower than copying a number of the same length. This assumption has since been confirmed in (Farrand, in prep.). The prediction assumptions are thus:

$$t_{KEY} = t_{PFCURSOR} = t_L = t_3 = \cdots$$
$$t_{CALC1(1)} > t_{COPY(1)}$$

We include the actual number to be calculated in parenthesis, rather than the length, for clarity. Specifically, we have,

$$t_{CALC1(295)} > t_{COPY(300)}$$

Then, substituting the first assumption in the two equations to be compared,

$$t_{locate} = 12t_{KEY} + t_{COPY(300)} + t_{CALC2(501)} \tag{3}$$
$$t_{scrollexact} = 12t_{KEY} + t_{CALC1(295)} + t_{CALC2(501)} \tag{4}$$

and substituting the second assumption in Eq. (4),

$$t_{scrollexact} = 12t_{KEY} + t_{COPY(300)} + t_{CALC2(501)} + C \tag{5}$$

From Eqs. (3) and (5), it is clear that the scroll exact strategy will take longer.

Prediction from Expanded Model

From the expanded (cognitive) model, we thus have the following prediction:

P1a. The scroll exact strategy will take longer than the locate strategy.

Notice that this is a *different* prediction than we obtained with the purely physical model. Notice also that this approach predicts a difference between the two user strategies—even though there are the same number of physical input actions in each *and* the same number of cognitive actions. The difference arises from the specific kinds of cognitive actions involved. We strongly suspect that this prediction is more likely to be correct. We are now testing an analogous prediction.

In the same way, we can make a prediction comparing CALC1 (here 295) and CALC2 here (501). Even though the length of both these strings is the same, and even though they both involve calculation, we predict that CALC2 will be harder than CALC1. (The underlying assumption is that a calculation which involves ordinary arithmetic ($300 - 5 = 295$) is easier than one which involves less straightforward arithmetic (subtract, then remember to add 1).

DISCUSSION

The introduction of cognition and the introduction of explicit prediction assumptions are both very powerful devices. Cognition is clearly central to ease of use. It can and should be represented explicitly in a description of a man–machine interface and in predictions about ease of use made from that description. The concept of "cognitive terminal symbol," furthermore, permits us to represent different kinds of cognitive actions (e.g., copy, count, calculate). We are thus in a position to compare the effect on ease of use of such actions with each other, and with physical input actions.

The prediction assumptions are also very powerful. They give us a much wider basis for making predictions than we had previously (string length and number of extra rules). They permit us to compare different kinds of physical actions with each other. They also permit us to compare physical actions with cognitive ones, and cognitive ones with each other. Furthermore, once some assumptions are stated, we can infer other assumptions from them for a still wider prediction base.

Furthermore, the prediction assumptions provide us with a great deal of flexibility in the prediction process. In particular, they permit us to make different predictions for different user populations. A frequent injunction to system designers concerned with ease of use is to "know the user." However, this injunction is not often followed by statements about precisely what is to be known about the user or about what is to be done with the resulting information. It is clear that knowledge of the user population is to be used *in creating the prediction assumptions.* For example, it is probable that for most user populations, the time to do mental arithmetic to calculate a number, as in the example above, will be longer than the time to enter that number at a computer terminal. However, for a user population who cannot type and are mathematicians, just the reverse might be true. With opposite assumptions, the predictions could also be opposite.

Specific prediction assumptions may not always be true. Even the assumption that typing a long string will take more time than typing a shorter one is not always correct. For example, typing a four-digit random number will probably take longer than typing a five-digit, frequently used word such as "these." By stating the assumptions concretely, however, it is then possible to refine them more precisely.

Another issue is the fact that the prediction assumptions are merely stated as equations or inequalities. Ideally, one would like explicit data from which to make numerical predictions instead of merely ordinal ones. Unfortunately, in many cases such data does not exist. In others, it is not easily available or not exactly what is needed. Some of the data desired does exist in some form in the literature. Keystroke times certainly exist. Mental arithmetic, too, is used as a task in a variety of psychological experiments. There have also been attempts to determine the mental processes involved in mental arithmetic which contain data (Aiken & Williams, 1973; Ashcraft & Stazyk, 1981), but the data is not exactly what is needed. Attempting to formulate the assumptions precisely, however, is still of value. We now know the information we do not have, and could obtain with behavioral science experimentation. Such information would not be restricted to the particular system being tested. Results of one experiment would probably be applicable to a number of different systems. Formulating prediction assumptions precisely, as in the above example, can lead us to either data or methodology *in the psychological literature.* It is generally felt that psychology has something to offer human factors. However, the relationship between the two is not entirely clear. Prediction assumptions, used in conjunction with a formal description of user actions may make one aspect of that relationship clearer.

The prediction assumptions can be made responsive to the level of analysis desired. Thus, an analysis which is primarily concerned with keyboard input might differentiate times to type different keys. However, an analysis which is concerned primarily with cognitive activities might consider time to type any key the same as time to type any other.

Although the methodology is beginning to look increasingly precise, the work is far from complete. Both the notions of cognitive terminal symbols and that of prediction assumptions are crucial. However, we have not yet given a precise method for finding either one, although we suspect that such methods can be developed. We suspect that there are a limited number of kinds of cognitive symbols and that these can be enumerated. This, however, is yet to be done. We further suspect that a limited number of rules can be specified for including cognitive terminal symbols in either a grammar or in sentences derived from a grammar. Such rules would be similar to our example, "if a number is to be typed, it must either be copied or calculated."

Explicit methods are also needed for obtaining the prediction assumptions. In the examples we gave, we generated the relevant sentences, then examined them to determine the prediction assumptions. It is to be hoped that the set of prediction assumptions can be "grown." That is, assumptions determined for one ease of use prediction will be sufficiently general to be used for other predictions.

Specific cognitive actions, other than those illustrated, must still be identified by human judgment, as must specific prediction assumptions. Human factors is still not a "predictive science." The procedure presented in this chapter is intended as one step toward that goal.

SUMMARY

We have presented an "action language" approach to predicting the ease of use of a man–machine interface. This method describes the users action language with a formal grammar which includes cognitive and physical input actions. It then generates sentences from the grammar for the comparisons to be made, converts the sentences to equations with time as variables, then substitutes "prediction assumptions" in the equations and solves the equations by simple computation.

Two concepts are crucial to this method. First, the concept of "cognitive terminal symbol" is introduced. This concept permits us to include cognition, which is central to ease of use, in the formal description. It permits us to represent different kinds of cognitive actions (e.g., count, calculate, copy) so that their effect on ease of use can be compared. Furthermore, by distinguishing "cognitve" from "physical input" symbols, we are closer to operationally definable terms than we were before. Such operational definition is a prerequisite for a tool to predict ease of use.

A second crucial concept is the notion of "prediction assumptions." The prediction assumptions embody, explicitly, the knowledge of experts in the field of human factors. These give us a much larger basis for making predictions than the two criteria used in previous work (string length and number of extra rules). While particular assumptions can be wrong, stating them explicitly (in the form of mathematical inequalities) permits them to be accepted, rejected or modified. Stating them explicitly also clarifies the kinds of data to be sought in either the human factors or the psychological literature, or in further experimentation. Furthermore, these prediction assumptions permit us to "tailor" predictions to the user population. What is hard for one kind of user will be easy for another. The prediction assumptions permit us to make different predictions about the same interface for different user populations.

The chapter gives detailed examples of the methodology. The main example compares different user strategies to identify a line in a text editor (using a locate command and scrolling with various options). Furthermore, it compares predictions about ease of use from a model with cognition and a model without it in order to determine whether inclusion of cognition makes any difference in the predictions obtained. From the physical input model alone, the method predicts that a "locate" strategy will take the same amount of time as a "scroll exact" strategy (entering the exact amount to be scrolled, then scrolling), and that a "scroll repeat" strategy (scrolling with a given default) will take longer than either. When cognition is added to the models, however, the predictions change. The new prediction is that the locate strategy will take less time than the scroll exact strategy—even though each has the same number of physical input actions and the same number of cognitive actions. Inclusion of cognition in the prediction process is therefore crucial to the predictions obtained.

The action language approach still requires human judgment to identify the cognitive actions and the prediction assumptions. With that disclaimer, the method is both explicit and powerful.

APPENDIX A. GRAMMAR FOR Dn

The following grammar gives a portion of a grammar for the Dn function (delete n lines) in a text editor. It gives the portion relevant to a comparison of different user strategies for identifying the first line to be deleted. Nonterminals are in small letters, terminal symbols in capitals. The numbering is intended to facilitate discussion and does not imply an ordering. The term NULL means no user action. Only physical input actions are described, not cognitive actions. In some cases, the words PRESS or TYPE are omitted when the meaning is clear

1. use Dn	→ identify first line + enter Dn command + PRESS ENTER

Possible User Strategies

2. identify first line	→ get first line on screen + move cursor to first line
3. get first line on screen	→ use locate strategy | use scroll strategy

Locate Strategy

4. use locate strategy	→ move cursor to command input field + type locate command + PRESS ENTER
5. move cursor to command input field	→ use cursor keys | PRESS PFCURSOR | NULL
6. type locate command	→ type locate keyword + type line number
7. type locate keyword	→ L+O+C | L | L+O+C+A+T+E
8. type line number	→ type number

"Scroll" Strategies

9. use scroll strategy	→ use scroll command + scroll with pfkey
10. use scroll command	→ use default scroll command | change scroll command
11. use default scroll command	→ NULL
12. change scroll command	→ move cursor to scroll field + type cursor parameter keyword

13. move cursor to scroll field	→	use cursor keys | PRESS PFCURSOR | PRESS PFCURSOR + PRESS PFCURSOR | NULL
14. type cursor parameter keyword	→	P+A+G+E | H+A+L+F+ | M+A+X | C+S+R | type parameter number | P | H | M | C
15. type parameter number	→	type number
16. scroll with pfkey	→	scroll key | scroll key + scroll with pfkey
17. scroll key	→	PRESS SCROLLUP | PRESS SCROLLDOWN

Miscellaneous Rules

18. enter Dn command	→	TYPE D + type numerical difference
19. type numerical difference	→	type number
20. type number	→	n | n + type number
21. n	→	1 | 2 | 3 | 4 | 5 | 6 | 7 | 8 | 9 | 0
22. move cursor to first line	→	use cursor keys | NULL
23. use cursor keys	→	indicate direction | indicate direction + use cursor keys | NULL
24. indicate direction	→	PRESS LEFTARROW | PRESS RIGHTARROW | PRESS UPARROW | PRESS DOWNARROW | PRESS CARRIAGERETURN

REFERENCES

Aiken, L. R., & Williams, E. N. Response times in adding and multiplying single-digit numbers. *Perceptual and Motor Skills*, 1973, *37*, 3–13.

Ashcraft, M. H., & Stazyk, E. H. Mental addition, a test of three verification models. *Memory and Cognition*, 1981, *9*, 185–196.

Bleser, T., & Foley, J. D. Toward specifying and evaluating the human factors of user-computer interfaces. *Proceedings, Human Factors in Computer Systems*. Gaithersburg, MD. March 1982, 309–314.

Branscomb, L. M. The human side of computers, *IBM Systems Journal*, 1981, *SE-7*, 229–240.

Card, S. K., & Moran, T. P. The keystroke-level model for user performance time with interactive systems. *Communications of the ACM*, 1980, *23*, 396–410.

Chomsky, N. *Syntactic structures*. The Hague: Mouton, 1964.

Embley, D. W., Empirical and formal language design applied to a unified control construct for interactive computing. *International Journal of Man–Machine Studies*, 1978, *10*, 197–216.

Embley, D. W., Lan, N. T., Leinbaugh, D. W., & Nagy, G. A procedure for predicting program editor performance from the users point of view. *International Journal of Man–Machine Studies*, 1978, *10*, 639–650.

Farrand, A. B. Keystrokes and cognition: an experimental comparison, in preparation.

Feyock, S. Transition diagram-based CAI/HELP systems, *International Journal of Man–Machine Studies*, 1977, *9*, 399–413.

Foley, J. D. The structure of interactive command languages. In R. A. Guedj et al. (Eds.), *Methodology of interaction*. Amsterdam: North-Holland Publishing Co., 1980.

Foley, J. D., & Wallace, V. L. The art of natural graphic man–machine conversation, *Proceedings of the IEEE*, 1974, *62*, 462–471.

Jacob, R. J. K. Using formal specification in the design of a human-computer interface, *Proceedings, human factors in computer systems*, Gaithersburg, MD, March 1982, 315–321.

Kieras, D. E., & Polson, P. G. An approach to the formal analysis of user complexity, *Project on the user complexity of devices and systems, Working Paper No. 2*. University of Arizona and University of Colorado, October 1982.

Ledgard, H. F., & Singer, A. Formal definition and design, *COINS Technical Report 78-01*. University of Massachusetts, Amherst, February, 1978.

Lewis, P. M., II, Rosenkranz, J., & Stearns, R. E. *Compiler Design Theory*. Reading, MA: Addison-Wesley, 1976.

McCormick, E. J. *Human factors engineering* (3rd ed.). New York: McGraw-Hill, 1970.

Moran, T. P. The command language grammar: a representation for the user interface of interactive computer systems. *International Journal of Man–Machine Studies*, 1981, *15*, 3–50.

Parnas, D. L. On the use of transition diagrams in the design of a user interface for an interactive computer system. *Proceedings of the 24th National ACM Conference*. New York: ACM, 1969, 379–385.

Reisner, P. Use of psychological experimentation as aid to development of a query language. *IEEE Transactions on Software Engineering*, 1977, *SE-3*, 218–229.

Reisner, P. Formal grammar and human factors design of an interactive graphics system, *IEEE Transactions on Software Engineering*, 1981, *SE-7*, 229–240.

Shaw, A. C. On the specification of graphics command languages and their processors. In R. A. Guedj et al. (Eds.), *Methodology of interaction*. Amsterdam, The Netherlands: North-Holland Publishing Co., 1980.

Schneiderman, B. Multiparty grammars and related features for defining interactive systems. *IEEE transactions on systems, man, and cybernetics*, 1982, *SMC-12*, 148–154.

Van Cott, H. P., & Kinkade, R. G. (Eds.). *Human engineering guide to equipment design*. Washington, DC: American Institutes for Research, 1972.

Wasserman, A. I., & Stinson, S. K. A specification method for interactive information systems, *Proceedings, conference on specifications of reliable software*, Cambridge, MA, 1979.

Woodson, W. E., & Conover, D. C. *Human engineering guide for equipment designers*. Berkeley, CA: University of California Press, 1973.

4

Stochastic Modeling of Individual Resource Consumption during the Programming Phase of Software Development*

DANIEL G. McNICHOLL
KENNETH MAGEL

In the past several years there has been a considerable amount of research effort devoted to developing models of individual resource consumption during the software development process. Since many conditions affect individual resource consumption during the software development process, including several which are difficult if not impossible to quantify, it is our contention that a stochastic model is more appropriate than a deterministic model.

In order to test our hypothesis, we conducted an experiment based upon several student programming assignments. Data from this experiment are used to demonstrate that the two parameter log-normal distribution is appropriate for describing the probabilistic behavior of the random variable "resource consumption." In addition we present a theoretical argument for the applicability of the log-normal distribution based on the concept of a proportional effects model for the growth of a program.

There is a consensus among software engineers (Faraguhar, 1970; Tonies, 1979; Zelkowitz et al., 1979) that one of the more important aspects of software development management is the prediction of resource consumption during the software development process. Given this impetus there has been a marked increase in research efforts during the past several years toward developing accurate predictive models of individual resource consumption with only moderate degrees of success. Almost without exception these efforts have used deterministic models of the software development process. We believe the underlying nature of software development is inherently stochastic. The success of a stochastic predictive model of software development will depend on two factors: (a) the adequacy of the distribution chosen to explain the probabilistic behavior of the process; and (b) the ability of the model to determine the parameters of the distribution for a particular development. Only the first of these two criteria will be addressed in this chapter.

*This work was supported in part by NSF Grant MCS 8002667.

Since we want to develop precise rather than qualitative results, we have concentrated on a single phase of the software development process, the Programming Phase. Although historically this phase has only accounted for approximately 10–20% of the total resources consumed during a software development process, it is a natural starting point since it is one of the most mechanical and therefore measurable phases of the process. Given the somewhat loose use of terminology within the software engineering discipline, we state here our working definition of the "programming phase" to avoid confusion:

> The PROGRAMMING PHASE of a software development process is identified as that task which: (a) has as its inputs a complete description of the developing program's inputs, outputs, and the functional dependencies of the output on the input; and (b) produces a correct program in a machine-readable language.

The immediate goal of our investigation was to determine the probabilistic behavior, as realized by a probability distribution, of various resource consumption quantities during the programming phase of software development. Our approach was to devise an experiment which would allow accurate measurements of all the relevant conditions and outcomes. From the collected data evidence is shown in the section entitled "Justification for the Stochastic Model" that justifies our premise that a stochastic model is needed. From the resultant experimental data we then tested the goodness-of-fit of various postulated theoretic distributions as described in the section after that. Equally important to us was the derivation of a theoretic rationale for the empirical results; this is presented in the penultimate section. In the final section our results are summarized and followup research is outlined.

DESCRIPTION OF THE EXPERIMENT

Sixty students in two sections (same instructor) of a college sophomore-level class on PL/I were given six written programming assignments. These students had taken two previous computer science courses using Fortran and one using assembly language. The assignments, which were relatively simple, contained a complete description of the program's inputs and outputs as well as the functional dependencies of the output on the input. The resources which the students consumed were collected via two mechanisms. During the students' attempts to solve the assignments certain data were automatically collected by the computer's accounting system. When processed, this data yielded three resource usage meters:

1. #RUNS: The total number of computer runs made.
2. CPU-TIME: The total cpu time, as measured in seconds, used to develop the program.
3. CPU COST: The total cpu charges, as measured in dollars, incurred during the assignment.

In addition, after the assignment was finished the students completed a questionnaire on each assignment from which the following resource usage meter was extracted.

#HOURS: The total number of hours of expended effort.

Information on the individual characteristics of the students was collected at the beginning of the course through a written survey. Tables 1,

TABLE 1
Descriptive Statistics of the Resource Usage Data[a]

Resource Meters	Program IDs	Sample Statistics		
		Size	Mean	Std. Dev.
#HOURS	TRI	56/53	8.0/ 7.1	7.2/ 5.7
	RKT	50/46	13.7/14.0	8.4/ 8.3
	INS	42/35	18.6/19.0	17.7/19.1
	SRT	44/42	10.8/11.1	10.0/10.2
	MST	37/34	16.3/16.5	15.2/15.9
	MML	40/32	21.6/22.3	19.0/21.0
#RUNS	TRI	59/56	13.9/13.7	7.4/ 7.4
	RKT	53/48	20.9/21.3	16.9/17.5
	INS	53/41	29.4/28.7	21.5/20.5
	SRT	52/48	21.1/21.6	18.1/18.4
	MST	48/40	24.5/25.2	17.5/18.2
	MML	48/37	28.0/26.6	21.0/15.8
CPU-TIME	TRI	59/56	3.6/ 3.5	1.9/ 1.8
	RKT	53/48	8.4/ 8.4	6.7/ 6.7
	INS	53/41	13.5/13.6	13.3/13.8
	SRT	52/48	13.0/13.6	13.2/13.4
	MST	48/40	22.4/23.5	22.4/23.9
	MML	48/37	21.3/20.3	19.5/12.8
CPU-COST	TRI	59/56	0.7/ 0.6	0.4/ 0.4
	RKT	53/48	1.9/ 2.0	1.8/ 1.8
	INS	53/41	3.0/ 3.0	2.5/ 2.5
	SRT	52/48	1.4/ 1.5	1.2/ 1.2
	MST	48/40	2.7/ 2.8	2.7/ 2.9
	MML	48/37	3.7/ 3.8	3.0/ 2.9

[a]Where two values are given the first value corresponds to the statistic for all students who attempted the assignment; the second applies to only those students who successfully completed the assignment.

TABLE 2
Descriptive Statistics of the Individual Characteristic Data[a]

Individual Characteristic	Sample Statistics			
	Size	Mean	Median	Std. Dev.
EDLEVEL	60	67.7	63	29.9
EDLOAD	60	14.9	15	2.86
EMPLOAD	60	4.17	0	10.5
CAMPEXP	60	3.82	3	3.26
DEPTEXP	60	3.57	3	2.54
WORKEXP	60	21.0	11	28.1
DPEXP	60	4.27	0	7.64
CGPA	59	2.95	2.9	0.6
AGE	60	22.4	21.1	3.75

	Distribution of Responses (%)					
	Very High	High	Good	Fair	Low	None
Interest	45.8	40.7	8.5	3.4	0.0	1.7
Ability	3.4	33.9	49.2	13.6	0.0	0.0
Enjoyment	22.0	44.1	20.3	13.6	0.0	0.0

[a]See Table 3 for definitions of individual characteristics.

TABLE 3
Definition of Individual Characteristics

Glossary of Individual Characteristics	
EDLEVEL	The number of college credit hours completed prior to the course
EDLOAD	The number of credit hours the student was taking simultaneously with the PL/I course
EMPLOAD	The number of hours per week that a student works at a job
CAMPEXP	The number of semesters completed at the campus
DEPTEXP	The number of CSC courses taken at the campus
WORKEXP	The number of full-time months, or equivalent for part-time, worked at any job since graduating from high school
DPEXP	Same as WORKEXP, except for DP-related jobs only
CGPA	Cumulative grade point average
Age	Student's age in years at beginning of the course
Interest	The student's rating of his interest in computer science
Ability	The student's rating of his ability as a programmer
Enjoyment	The student's rating of his enjoyment of programming

TABLE 4
Descriptive Statistics of the Number of Statements Program Size Meter (Includes All Types of PL/I Statements)

Program	Mean	Std. Dev.	Low	Medium	High
TRI	22	8.0	14	20	71
RKT	68	15.5	49	66	134
INS	73	16.5	51	70	127
SRT	47	10.8	34	45	96
MST	85	15.1	61	82	125
MML	116	24.9	73	112	181

2, and 3 present some simple univariate descriptive statistics of the data collected from the experiment. The number of PL/I statements in the resultant programs are summarized in Table 4 to give some idea of the size of each assignment.

JUSTIFICATION FOR THE STOCHASTIC MODEL

If we assume for the moment that there exists a deterministic model of individual resource consumption during the programming phase, then there must exist a function g_x such that:

$$R_x = g_x(Z_1, Z_2, ..., Z_n),$$

where R_x is the amount of resource X consumed by a specific individual on a specific programming assignment and $Z_1, Z_2, ..., Z_n$ are meters of all the relevant sources of variation in the programming process. For example, programmer, program assignment, environment.

In the experiment described in the previous section there are only three potential sources of variation: the programming assignment, the programmer, and the sequence of assignments. We might then hypothesize that there must exist a function g_x such that:

$$R_{x/ij} = g_x(I_i, P_j, S_j),$$

where $R_{x/ij}$ is the amount of resource X consumed by individual i on programming assignment j, I_i is some set of characteristic measurements for individual i which account for the variation among individuals' resource usage, P_j is some set of complexity/size metrics for programming assignment j which will account for the variation of resource usage between assignments, and S_j is the sequence of programming assignment j which possibly could account for some additional interprogram variation.

Contrasted to the deterministic model represented by Eq. (1) is a stochastic model in which we hypothesize that there are so many un-measurable conditions affecting an individual's resource usage that we can not completely determine it. Therefore, we settle for describing the probabilistic behavior of the resource consumption in terms of its proba-bility density (p.d.f.) or cumulative distribution functions (c.d.f.), e.g.,

$$R_{x/j} \sim f(r_{x/j}; \theta_{x/j1}, \theta_{x/j2}, \ldots, \theta_{x/jn}) \quad 2$$

where $R_{x/j}$ is a random variable defined as the amount of resource X consumed by any individual on programming assignment j, f is the prob-ability density function of resource usage, and $\theta_{x/j1}, \theta_{x/j2}, \ldots, \theta_{x/jn}$ are parameters of the p.d.f. for resource X and programming assignment j.

Furthermore, we hypothesize that the values of the parameters are predictable from some measurements of all the nonindividual sources of variations. In our experiment the mathematical model for the param-eters would be:

$$\theta_{x/jk} = h_{x/k}(P_j, S_j) \quad 3$$

There exists a function h for each parameter of each resource consump-tion random variable distribution which can be determined from the given assignment and order.

To recap, the deterministic model of Eq. (1) assumes that we can measure all the relevant individual characteristics which will allow us to predict the actual resource consumption of that individual on a pro-gramming assignment. The stochastic model of Eqs. (2) and (3) assumes that there is no possibility of measuring all the relevant individual char-acteristics and therefore we must build our model such that when given all the relevant assignment characteristics it will be able to describe the probabilistic behavior of the resource consumption. In the stochastic model we are not able to predict individual resource usage but we can predict such statistics as the expected values, confidence ranges, for example.

To determine whether a deterministic model is appropriate we need to verify whether or not there exists individual characteristics which account for the interindividual variation of resource usage. If we assume for the moment that a deterministic model does exist, we can propose the model:

where $$\Phi_{x/ij} = g(I_i), \quad 4$$

$$\Phi_{x/ij} = [R_{x/ij} - \mu(R_{x/j})]/\sigma(R_{x/j})$$

Here, $\Phi_{x/ij}$ is the standardized amount of resource X consumed by indi-vidual i on program j. In other words $\Phi_{x/ij}$ is the number of standard deviations in terms of resource X for individual i on program j. The expression $\mu(R_{x/j})$ is the mean amount of resource X consumed by all students on program j; the expression $\sigma(R_{x/j})$ is the standard deviation of the resource X consumption on program j, and g is a function which when given the appropriate individual characteristic(s) will yield the individual's standardized amount of resource X on assignment j.

The validity of this model depends on two assumptions. The first is that an individual's standardized resource usage is independent of the program meters, i.e., if an individual is productive for large programs he will be for small ones and vice versa. The second assumption is that either the individual characteristics do not change with time, or they change relatively uniformly for all individuals.

Using the date that was collected on individual characteristics (see Table 1–4), a series of linear correlation analyses were performed using the model in Eq. (4). The results of these correlation analyses show that none of the individual characteristics measured in this experiment account for much of the interindividual variation of resource usage (see Table 5). Other researchers including Chrysler (1978) have also indicated difficulty in determining individual characteristic meters which are predictive of individual resource consumption. Although it has been shown by DeNelsky and McKee (1974) and by McNamara and Huges (1961) that certain programmer's aptitude tests are moderately predictive of job performance as measured by supervisory ratings, there has been no evidence that these tests predict individual resource variations.

An argument could be made that the failure to find an adequate

TABLE 5
Results of the Correlation Analysis of the Individual Characteristics with the Standardized Resource Usage Variables[a]

Individual Characteristics	Resource Usage Meters			
	#Hours	#Runs	CPU-time	CPU-cost
EDLEVEL	.001	.000	.000	.000
EDLOAD	.000	.000	.001	.007
EMPLOAD	.032	.000	.000	.000
CAMPEXP	.027	.007	.002	.002
DEPTEXP	.144	.004	.000	.000
WORKEXP	.002	.013	.008	.011
DPEXP	.002	.002	.003	.007
CGPA	.078	.173	.108	.134
AGE	.001	.004	.001	.003
INTEREST	.030	.001	.002	.012
ABILITY	.121	.003	.003	.003
ENJOY	.018	.017	.024	.020

[a]Results are in terms of the coefficient of determination (R^2).

TABLE 6
Results of the Canonical Correlation of Individual Characteristics as a Group with the Standardized Resource Usage Variables[a]

Resource Meter	R^2
#HOURS	.256
#RUNS	.276
CPU-COST	.247
CPU-TIME	.185

[a]Results are in terms of the coefficient of determination.

measure of individual resource variation does not imply that one does not exist. However, when a canonical regression analysis is performed to determine the ability of the individual characteristics as a group to explain the interindividual variation they account for little of the variation, (see Table 6). In addition, if we examine the ranges[1] of the standardized resource consumption amounts, $\Phi_{x/ij}$, of the data collected from our experiment most individuals vary significantly in their standardized resource consumption (see Figure 1 and Table 7).

In previous research (McNicholl & Magel, 1982) the authors have shown that the perceived complexity of a programming task is a highly subjective entity and thus the deterministic prediction of individual resource consumption depends upon understanding the psychology of the individual—a formidable, if not impossible task. All these points would seem to weigh against the possibility of finding a set of individual characteristics adequate for the deterministic model.

It therefore seems to us that the assumption of a deterministic model is not supportable. In the next section we shall examine the validity of the stochastic model.

DETERMINATION OF THE RESOURCE CONSUMPTION DISTRIBUTION

Given the premise that the programming task of a software development process is essentially a stochastic process, it becomes necessary to define the resource consumption quantities as random variables. In the

[1]The range for an individual is defined to be the absolute difference between the maximum and minimum $\Phi_{x/ij}$ for all j.

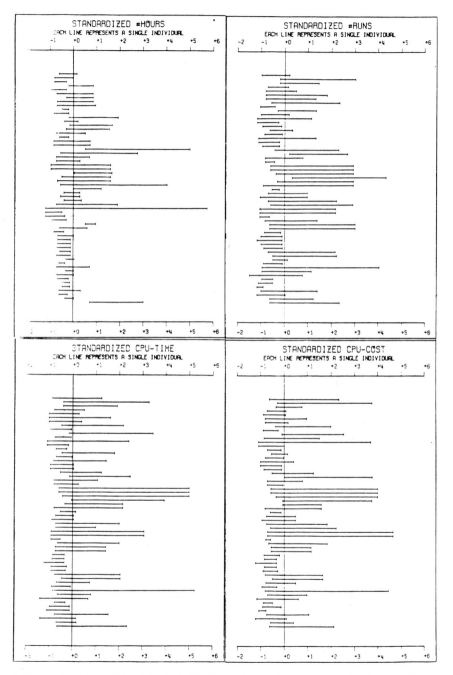

FIGURE 1. **Deviation plots of the individuals' standardized resource consumption variables.**

TABLE 7
Distribution of Individuals' Standardized Resource Usage Ranges

Ranges	Frequency (%)	Cumulative Frequency (%)
	Standardized # Runs	
0.00–0.25	0.0	0.0
0.25–0.50	8.5	8.5
0.50–1.00	25.4	33.9
1.00–2.00	16.9	50.8
2.00–3.00	27.1	78.0
3.00–4.00	18.6	96.6
4.00–5.00	1.7	98.3
5.00–6.00	1.7	100.0
6.00–7.00	0.0	100.0
	Standardized # Hours	
0.00–0.25	0.0	0.0
0.25–0.50	13.6	13.6
0.50–1.00	37.3	50.9
1.00–2.00	28.8	79.7
2.00–3.00	6.8	86.4
3.00–4.00	8.5	94.9
4.00–5.00	3.4	98.3
5.00–6.00	0.0	98.3
6.00–7.00	1.7	100.0
	Standardized CPU-COST	
0.00–0.25	1.7	1.7
0.25–0.50	5.1	6.8
0.50–1.00	30.5	37.3
1.00–2.00	27.1	64.4
2.00–3.00	18.6	83.1
3.00–4.00	3.4	86.4
4.00–5.00	8.5	94.9
5.00–6.00	5.1	100.0
6.00–7.00	0.0	100.0
	Standardized CPU-TIME	
0.00–0.25	0.0	0.0
0.25–0.50	5.1	5.1
0.50–1.00	25.4	30.5
1.00–2.00	22.0	52.5
2.00–3.00	30.5	83.1
3.00–4.00	10.2	93.2
4.00–5.00	0.0	93.2
5.00–6.00	5.1	98.3
6.00–7.00	1.7	100.0

experiment described in the previous section let the observable outcomes of interest be designated as R_h, R_r, R_f, and R_c and defined as the resource consumption of a successfully completed programming task as measured in man-hours, computer-runs, cpu-seconds, and cpu-dollars, respectively. In the following discussion the abstract random variable R will be used to denote any of these specific random variables.

Since the distinguishing property of a random variable is the probability value associated with each event of a measurement of that random variable, the probabilistic behavior of the outcomes can be completely described by identifying their probability distributions.

In order to derive a set of postulated theoretic distributions which might describe the empirical data we examined the histograms of R_h R_r, R_t, and R_c. A representative sample of these histograms is given in Figure 2. It is readily apparent that the histograms all display a general unimodal shape and positive skewness. Five theoretic distributions were selected as postulates based on their ability to assume the desired shape and skewness. These distributions were the log-normal, beta, gamma, Weibull, and type I extreme value maxima. In addition the normal distribution was used for comparison. The exact forms of the distributions selected are shown in Appendix A.

In order to investigate the adequacy o the six postulated distributions to explain the empirical distributions of the collected data, a series of chi-square goodness-of-fit tests were performed. The methods used to estimate the parameter values of the theoretic distributions from the sample are given in Appendix A. Figure 3 depicts the six theoretic p.d.f.'s overlaid on a sample resource histogram. The results of the chi-square tests are shown in Table 8.

The log-normal distribution best fits the empirical data in most instances. Possible reasons for the appropriateness of this distribution will be examined in the next section.

Although the foregoing goodness-of-fit tests were performed correctly there was one major, but intentional, omission. In the conduct of the experiment it was likely that some individuals would either not complete an assignment or complete it unsuccessfully. In either case these individuals were excluded from the samples of R_h R_r, R_t and R_c in the previous goodness-of-fit analysis because they did not meet the definition of the random variables—i.e., they were not "successfully completed." The presence of this multiple-random censoring causes complications because the censored data do not provide complete information, i.e., tell us when the task was successfully completed. Yet the censored data do provide partial information which is that up to the point of censoring successful completion did not take place. To incorporate this partial information available from the multiple censored data, and therefore

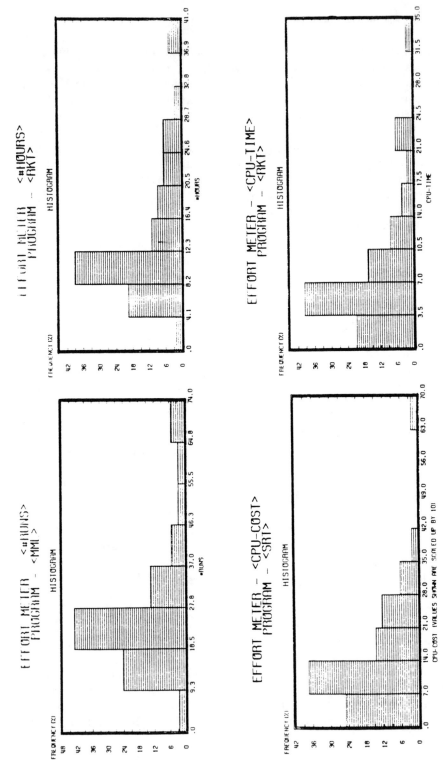

FIGURE 2. Sample histograms of the resource consumption random variables.

construct a more accurate model, we can make use of a procedure[2] outlined by Bury (1975).

Let us define a function of each of the resource consumption random variables called the "Completion Intensity Function" or the "Instantaneous Rate of Completion" and designate it as $v(r)$[3]. This function will yield the conditional probability of successfully completing the programming task with resource consumption $R = r$ given that $R \geq r$. Let us define $f(r)$ as the probability density function (p.d.f.) of R and $F(r)$ as the cumulative distribution function (c.d.f.) of R. Clearly $f(r) = P(R = r)$ and $F(r) = P(R \leq r)$, and therefore the instantaneous rate of completion can be obtained by

$$v(r) = f(r)/[1 - F(r)]$$

To obtain the actual instantaneous rates of completion from the empirical data we will use the following. Let the measurement domain of the random variable R be divided into adjoining intervals $\Delta r_1, \Delta r_2, ..., \Delta r_n$. Denote the number of occurrences in Δr_i as $c(r_i)$. The relative frequency of occurrences of measurements in Δr_i, denoted by $\eta(r_i)$, is then the ratio of the number of occurrences to the sample size, i.e., $\eta(r_i) = c(r_i)/n$; where n includes both completed and censored data points. Since theoretically it is possible to obtain larger and larger samples while decreasing the interval size such that $n * \Delta r_i$ remains finite, then the relative frequency of observations per interval approaches the probability density of R. That is,

$$\lim_{\substack{n \to \infty \\ \Delta r_i \to dr}} \eta(r_i)/\Delta r_i = f(r_i)$$

A like argument can be followed to show that

$$\lim_{\substack{n \to \infty \\ \Delta r_i \to dr}} \sum_{j=1}^{i} \eta(r_j) = F(r_i)$$

We can now restate the formula for the instantaneous rate of completion using the above results as

$$v(r_i) = \lim_{\substack{n \to \infty \\ \Delta r_i \to dr}} [c(r_i)/(n \times \Delta r_i)]/[1 - \sum_{j=1}^{i} c(r_j)/n]$$

[2]The applicability of this and other procedures from the area of reliability theory is easily understood if one considers the parallel nature of that theory and our research. The principal unknown random variable in reliability theory is TIME TO FAILURE, in our research the unknown random variable is similar but opposite, RESOURCE USAGE TO COMPLETION.

[3]Remember that the random variable R is an abstract one representing the actual random variables R_h, R_r, R_t, and R_c.

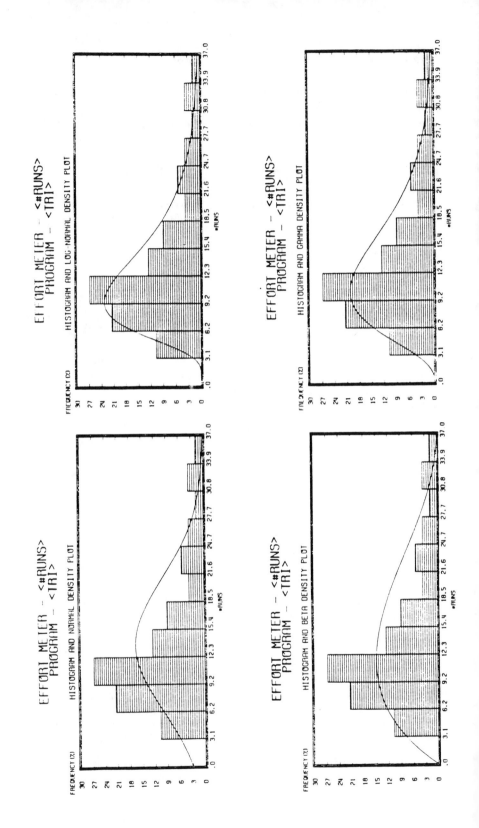

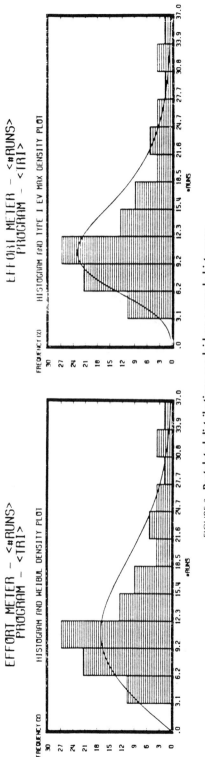

FIGURE 3. Postulated distributions overlaid on a sample histogram.

93

TABLE 8
Results of the χ^2 Goodness-of-fit Tests for the Six Postulated Distributions with the Resource Usage Data (not including censored data points)[a]

PGM IDS	Postulated Distributions						df[b]	Tab χ^{2c}
	Norm	Ln	Beta	Gam	Weib	Max		
Effort Meter, # Runs								
TRI	33.6	29.3	24.1	**13.5**	29.6	17.5	8	15.5
RKT	31.1	**4.9**	19.1	18.0	16.9	13.5	6	12.6
INS	20.5	**0.9**	**9.1**	**3.3**	**6.8**	**2.9**	5	11.1
SRT	27.7	**2.6**	22.5	**6.0**	14.2	**6.7**	6	12.6
MST	30.0	**10.8**	14.8	**8.0**	**5.6**	**8.0**	5	11.1
MML	18.1	9.7	15.4	12.0	12.8	**7.5**	4	9.5
Effort Meter, # Hours								
TRI	56.6	20.8	29.8	18.5	44.2	20.0	7	14.1
RKT	33.8	**5.7**	31.1	**11.5**	17.0	19.3	6	12.6
INS	24.8	**4.4**	25.2	15.6	17.6	**7.6**	4	9.5
SRT	22.0	**9.4**	21.6	12.9	15.9	12.1	5	11.1
MST	25.3	**5.9**	26.7	13.3	11.2	11.5	3	7.8
MML	28.4	8.5	24.6	13.4	13.4	11.5	3	7.8
Effort Meter, Cpu-Time								
TRI	27.7	**8.4**	28.9	19.8	27.7	**14.7**	8	15.5
RKT	29.2	**10.9**	22.9	13.5	15.0	16.1	6	12.6
INS	29.8	**2.1**	18.5	**2.5**	**7.6**	**4.5**	5	11.1
SRT	36.0	**3.7**	23.2	**9.7**	**6.3**	11.6	6	12.6
MST	21.6	**5.2**	16.8	**9.2**	12.8	**7.2**	5	11.1
MML	25.2	14.3	33.2	12.8	17.3	11.2	4	9.5
Effort Meter, Cpu-cost								
TRI	29.2	**10.0**	28.9	18.2	21.0	**13.9**	8	15.5
RKT	55.5	**6.4**	27.0	**10.1**	16.9	16.1	6	12.6
INS	24.4	**2.9**	**9.1**	**4.5**	**5.2**	**8.0**	5	11.1
SRT	19.5	**4.1**	18.0	**6.0**	**9.0**	**4.9**	6	12.6
MST	16.0	**7.2**	20.0	11.2	18.0	**8.0**	5	11.1
MML	26.8	9.7	23.4	12.8	20.3	10.1	4	9.5

[a]Results are in terms of the calculated χ^2 values.
[b]*df*: Degrees of freedom.
[c]TAB χ^2: Tabulated χ^2 values at the .05 significancy level.
[d]Boldface indicates that calculated χ^2 value shows that the distribution is not rejected at the .05 significance level.

If we now order all the observations according to the amount of resource R they consumed at completion or censoring, we can then define the "cumulative" instantaneous rate of completion as

$$V(r_i) = \sum_{j=1}^{i} v(r_j) \times \Delta r_j$$

or

$$V(r_i) = \lim_{\substack{n \to \infty \\ \Delta r_i \to dr}} \sum_{j=1}^{i} [c(r_j)/(n - \sum_{k=1}^{j} c(r_k))] \tag{5}$$

The denominator of the above equation now represents the total number of noncompletions just prior to the resource consumption r_i, including those data that were subsequently censored.

Since we desire the distribution of R, i.e., $F(r)$, we need to establish a relationship between it and $V(r)$. By definition we know that

$$v(r) = f(r)/[1 - F(r)]$$

by a simple transformation we obtain the following:

$$v(r) = [d(F(r))/dr]/[1 - F(r)]$$

and thus

$$v(r)dr = d[-\ln(1 - F(r))]$$

integrating the above from a truncation point r_0 to r yields

$$\int_{r_0}^{r} v(r)\ dr = -\ln[1 - F(r)]\Big|_{r_0}^{r}$$

Since $F(r_0) = 0$

$$\int_{r_0}^{r} v(r)\ dr = -\ln[1 - F(r)]$$

Hence

$$F(r) = 1 - \exp\left(-\int_{r_0}^{r} v(r)\ dr\right)$$

And since by definition

$$V(r) = \int_{r_0}^{r} v(r)\ dr$$

then

$$F(r) = 1 - \exp[-V(r)] \tag{6}$$

Summarizing the above we now know that we can develop an empirical c.d.f. using all the available information from the sample by means of Eq. (5) and (6). It is now possible to test the postulated distributions' c.d.f.'s for their adequacy in explaining this empirical c.d.f.

A series of nonlinear regression analyses were performed to determine the adequacy of the six postulated distributions' c.d.f.'s to explain the empirical c.d.f.'s of the experimental data formed from Eqs. (5) and (6). The results of these analyses are shown in Table 9. Sample plots of the theoretic and empirical c.d.f.'s are displayed in Figure 4. Once again

TABLE 9
**Results of the Nonlinear Regression Analyses of the Postulated Distributions' C.D.F.'s
Ability to Describe the Empirically Derived C.D.F.**[a]

Effort Meters	PGM IDS	Postulated Distributions					
		Norm	Ln	Beta	Gam	Weib	Max
#RUNS	TRI	.072	.037	.063	.053	.057	.049
	RKT	.074	.029	.051	.042	.049	.051
	INS	.066	.018	.053	.101	.034	.044
	SRT	.064	.020	.041	.032	.038	.041
	MST	.059	.037	.040	.046	.035	.043
	MML	.079	.047	.065	.070	.068	.059
#HOURS	TRI	.071	.033	.052	.072	.052	.052
	RKT	.079	.041	.066	.053	.062	.056
	INS	.083	.035	.058	.064	.055	.062
	SRT	.057	.026	.040	.079	.041	.041
	MST	.043	.040	.037	.093	.038	.039
	MML	.091	.056	.076	.069	.076	.074
CPU-TIME	TRI	.073	.040	.063	.051	.062	.049
	RKT	.073	.028	.051	.074	.048	.150
	INS	.065	.019	.036	.065	.036	.043
	SRT	.054	.021	.031	.083	.032	.035
	MST	.053	.035	.037	.080	.034	.042
	MML	.096	.060	.076	.071	.078	.074
CPU-COST	TRI	.069	.040	.055	.155	.060	.323
	RKT	.067	.024	.045	.075	.045	.046
	INS	.064	.021	.041	.067	.039	.042
	SRT	.060	.023	.036	.090	.035	.038
	MST	.049	.043	.037	.058	.037	.039
	MML	.070	.049	.063	.057	.066	.056

[a]Results are in terms of the standard error of the estimate (s.e.e.).

the log-normal distribution appears to best explain the empirical data. The values of the parameters of the log-normal distributions as determined by these regression analyses are presented in Table 10.

To statistically verify the significance of the theoretic log-normal c.d.f., a series of Komolgorov goodness-of-fit tests were performed using the parameter values from Table 10. The results of these tests are shown in Table 11 and demonstrate that the log-normal distribution is statistically significant as a theoretic distribution for the empirical data.

Up to this point we have seen only empirical justification for the applicability of the log-normal distribution to the resource consumption during the programming phase. In the next section we will present both an informal and a formal argument to explain why this applicability might exist.

RATIONALE FOR THE LOG-NORMAL DISTRIBUTION

In the previous section we have shown by empirical analyses that the log-normal theoretic distribution is a reasonable one to adopt in our attempt to describe the probabilistic behavior of individual resource consumption during the programming phase of a software development process. Shortly, we will present a formal theoretic argument for the applicability of this distribution, but first it will be insightful to examine informally the characteristics of the postulated distributions to determine whether they match expectations based on our intuitive understanding of the programming process.

From our knowledge of the programming process we know that there must exist a lower bound to the resource consumption random variable at $R = 0$. In order for distributions such as the normal or Type I extreme value maxima, which theoretically have no lower bounds, to be applied to our process we must impose an artificial truncation below the point $R = 0$. However, the log-normal, beta, gamma, and Weibull distributions do have theoretic lower bounds at the appropriate point making them more appropriate then those that do not.

We also know that there does not exist an upper bound to resource consumption, at least theoretically. Of the six postulated theoretic distributions, all but the beta distribution have no upper bounds.

Intuitively, we would expect the distribution of individual resource consumption to be such that there are a small number of highly productive individuals, i.e., individuals who use less resources, a larger number of individuals who are moderately productive, and a small number of individual who display low productivity. In other words we would expect the distribution to be unimodal. All six of the postulated distributions possess the ability to assume a unimodal shape given appropriate values for their parameters.

Another facet we can examine is the nature of the completion intensity function. We would expect that as time or any other resource quantity is expended the chances of an individual completing an assignment correctly, given that he or she has not already completed it, would increase—up to a point. At some point in the resource spectrum it seems natural that this conditional probability should start to decrease due to such factors as falling motivation, getting "lost" on the assignment, etc. Of all the postulated distributions only the log-normal has a completion intensity function which is nonmonotonically increasing. Figure 5 shows some sample Completion Intensity function plots based on our experimental data for the six postulated distributions.

Table 12 recaps the preceding discussion and demonstrates that, as far as intuition is concerned, the log-normal distribution once again

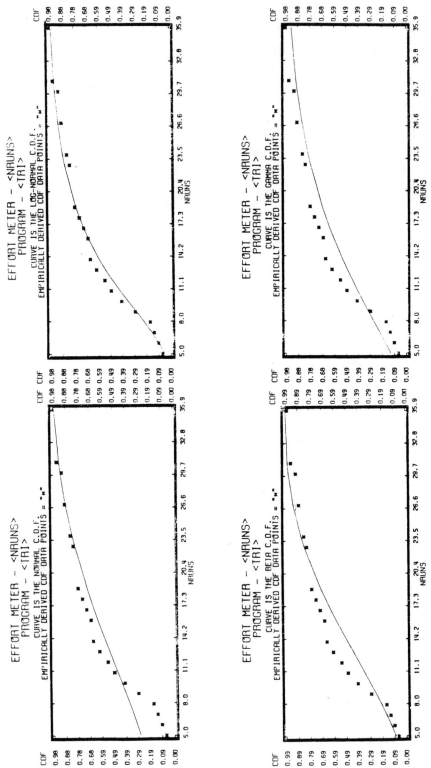

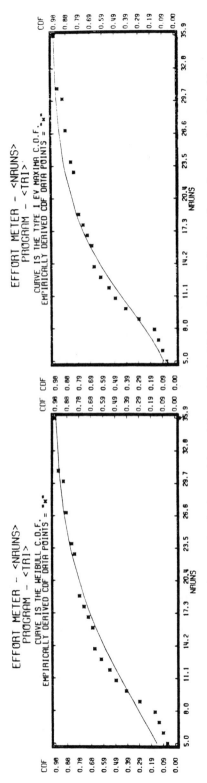

FIGURE 4. Sample plots of the theoretic distributions' c.d.f.'s and the empirical c.d.f. formed from all the data; theoretic parameters derived from the nonlinear regressions analyses.

99

TABLE 10
Parametric Results of Log-Normal Nonlinear Regression Analysis of the Empirical C.D.F. Formed from All Data

Effort Meter	PGM IDS	θ_1	θ_1 Asymptomatic Std. Error	θ_2	θ_2 Asymptomatic Std. Error
#RUNS	TRI	2.483	.017	0.578	.026
	RKT	2.831	.015	0.759	.025
	INS	3.285	.009	0.867	.016
	SRT	2.795	.009	0.780	.016
	MST	3.088	.021	0.863	.034
	MML	3.224	.019	0.603	.036
#HOURS	TRI	1.759	.024	0.720	.038
	RKT	2.468	.019	0.613	.031
	INS	2.683	.022	0.834	.038
	SRT	2.105	.016	0.736	.027
	MST	2.581	.018	0.504	.028
	MML	2.866	.031	0.731	.056
CPU-COST	TRI	−.584	.011	0.516	.019
	RKT	0.397	.009	0.745	.015
	INS	0.979	.009	0.793	.016
	SRT	0.146	.009	0.775	.016
	MST	0.795	.021	0.897	.035
	MML	1.193	.015	0.549	.029
CPU-TIME	TRI	1.150	.012	0.516	.020
	RKT	1.891	.011	0.782	.020
	INS	2.436	.008	0.827	.015
	SRT	2.293	.008	0.802	.014
	MST	2.867	.019	1.057	.035
	MML	2.906	.022	0.661	.042

seems appropriate. While intuition alone is often misleading, in this case it is supported by the empirical evidence.

It is appropriate to mention at this point that prior to our data analyses the authors did not expect the log-normal to be the most suitable of the postulated distributions. When the empirical data convinced us that it was, we felt it necessary to attempt a theoretic argument for the log-normal distribution in order to better understand its applicability. Once again we drew upon the discipline of reliability theory to develop a possible theoretic argument for the applicability of the log-normal distribution. The following argument closely follows that of Mann, Schafer, and Singpurwally (1974) where it was applied to the applicability of the log-normal to a fatigue-life model.

The central theme of our argument may be stated informally as relying on a "proportional effects model" for the "growth" of a program.

TABLE 11
Results of the Komolgorov Goodness-of-fit Tests of the Log-Normal
Distribution to the Empirical C.D.F. Formed from All the Data
(In all cases the hypothesis was not rejected at the
.05 significance level)

Effort Meters	PGM IDS	KCAL[a]	N[b]	KTAB[c]
#HOURS	TRI	.076	53	.187
	RKT	.076	46	.200
	INS	.110	35	.224
	SRT	.065	42	.210
	MST	.099	34	.227
	MML	.160	32	.234
#RUNS	TRI	.077	56	.183
	RKT	.073	48	.196
	INS	.066	41	.212
	SRT	.049	48	.196
	MST	.073	40	.210
	MML	.129	37	.218
CPU-COST	TRI	.073	56	.183
	RKT	.055	48	.196
	INS	.051	41	.212
	SRT	.056	48	.196
	MST	.088	40	.210
	MML	.120	37	.218
CPU-TIME	TRI	.073	56	.183
	RKT	.068	48	.196
	INS	.070	41	.212
	SRT	.052	48	.196
	MST	.063	40	.210
	MML	.123	37	.218

[a] KCAL: calculated K statistic.
[b] N: sample size.
[c] KTAB: tabulated K statistic at the appropriate degrees of freedom based on
N and at the .05 significancy level.

This implies that the growth of a program at each "stage" of its development is randomly proportional to its degree of "maturation" at the previous stage. The program starts out at some minute, but nonzero, degree of maturation and continues to mature until it reaches some predefined completion degree of maturation. The following is a more formal expression of this argument.

Let $M_0 < M_1 < \cdots < M_n$ be a sequence of random variables that denote the degree of maturation of the program at successive stages of growth. The degree of maturation of a program can be thought of as its percentage of completion. A stage of growth can be viewed in micro terms using

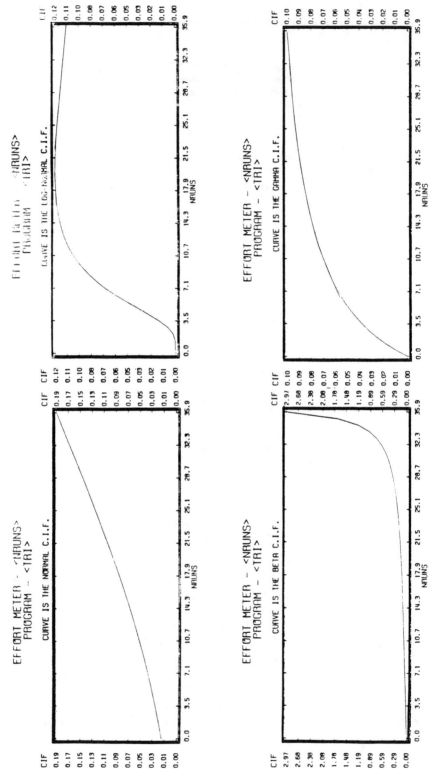

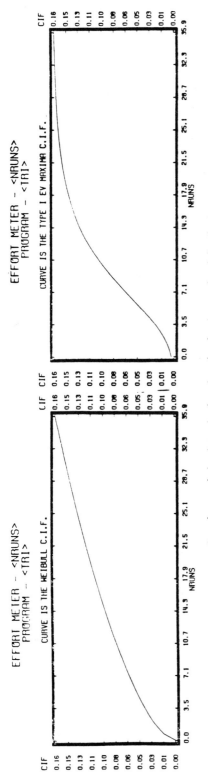

FIGURE 5. Sample completion intensity functions for the six postulated distributions.

103

TABLE 11
**Results of the Komolgorov Goodness-of-fit Tests of the Log-Normal
Distribution to the Empirical C.D.F. Formed from All the Data
(In all cases the hypothesis was not rejected at the
.05 significance level)**

Effort Meters	PGM IDS	KCAL[a]	N[b]	KTAB[c]
#HOURS	TRI	.076	53	.187
	RKT	.076	46	.200
	INS	.110	35	.224
	SRT	.065	42	.210
	MST	.099	34	.227
	MML	.160	32	.234
#RUNS	TRI	.077	56	.183
	RKT	.073	48	.196
	INS	.066	41	.212
	SRT	.049	48	.196
	MST	.073	40	.210
	MML	.129	37	.218
CPU-COST	TRI	.073	56	.183
	RKT	.055	48	.196
	INS	.051	41	.212
	SRT	.056	48	.196
	MST	.088	40	.210
	MML	.120	37	.218
CPU-TIME	TRI	.073	56	.183
	RKT	.068	48	.196
	INS	.070	41	.212
	SRT	.052	48	.196
	MST	.063	40	.210
	MML	.123	37	.218

[a]KCAL: calculated K statistic.
[b]N: sample size.
[c]KTAB: tabulated K statistic at the appropriate degrees of freedom based on N and at the .05 significancy level.

the principle of cognitive psychology that information is processed in stages (Curtis, 1981).

Let ΔM_i be the increment that the maturity of the program was increased by in stage i, i.e., $\Delta M_i = M_i - M_{i-1}$. Given the proportional effects model we know

$$\Delta M_i = \rho_i M_{i-1},$$

where ρ_i, the constant of proportionality, is a random variable assumed to be independent of all other ρ_j's, but not necessarily with the same distribution.

TABLE 12
Comparison of Postulated Distributions' Characteristics with the "Ideal"

Postulated Distributions	*Conditions[a]*		
	I	*II*	*III*
Normal	No	Yes	No
Log-Normal	Yes	Yes	Yes
Beta	Yes	No	No
Gamma	Yes	Yes	No
Type I Extreme Value Maxima	No	Yes	No
Weibull	Yes	Yes	No
Ideal	Yes	Yes	Yes

[a]Conditions:
  I   Distribution has a theoretic lower limit at R = 0.
  II Distribution has no theoretic upper limit.
  III Distribution has a nonmonotonically increasing Completion Intensity Function.

Let M_0 be the program's degree of maturation at the beginning which is some minute but nonzero value. This can be thought of as the conceptual "seed" from which the program grows.

Let M_n be some predefined degree of maturation which has been specified to be the acceptable minimum maturity of the completed program.

Then

$$\rho_i = [\Delta M_i/M_{i-1}], \qquad i = 1, 2, \ldots, n$$

and thus:

$$\sum_{i=0}^{n} \rho_i = \sum_{i=0}^{n} [\Delta M_i/M_{i=1}]$$

Given that ΔM_i can get smaller and smaller while n increases such that $\Delta M_i n$ remains finite then:

$$\lim_{\substack{\Delta M_i \to 0 \\ n \to \infty}} \sum_{i=0}^{n} \rho_i = \int_{M_0}^{M_n} M^{-1} \, dM = \ln(M_n) - \ln(M_0)$$

or

$$\ln(M_n) = \sum_{i=1}^{n} \rho_i + \ln(M_0)$$

Since by assumption the ρ_i's are independent random variables it follows from the central limit theorem that

$$\sum_{i=1}^{n} \rho_i \sim \text{Normal } (\mu_1, \sigma_1{}^2)$$

and thus

$$\ln(M_n) \sim \text{Normal } (\mu_2, \sigma_2{}^2)$$

therefore,

$$M_n \sim \ln(\mu_2, \sigma_2{}^2)$$

If we assume that the resource consumption, R_i, at any stage is a function of the work performed, i.e.,

$$R_i = h(\Delta M_i)$$

and further assume that the function h is distribution preserving then:

$$R_n \sim \ln(\Theta_1, \Theta_2)$$

In summary, we have argued that if we assume a proportional effects model for the growth of a program and in addition assume that resource usage is a distribution preserving function of the program growth, then we reach the conclusion that the resource consumption random variable is log-normal. In this argument we have had to make a number of assumptions which to the authors seem reasonable.

CONCLUSIONS

From the preceding analyses we believe that the following conclusions are justified:

- Individual resource consumption is an outcome of a process which is inherently stochastic and should be viewed as a random variable rather than deterministically.
- The log-normal distribution appears to be an appropriate model to adopt in order to empirically describe the probabilistic behavior of individual resource consumption during the programming phase of software development.
- The log-normal distribution possesses characteristics which seem appropriate for describing resource usage during the programming process and in addition a theoretic argument can be constructed for it which depends on reasonable assumptions.

In addition to the data collected for the experiment described on pages 80–83 we have collected data from two other identical experiments which we are currently processing to cross-validate the log-normal model.

As with many of the experiments conducted within the area of software engineering, our research is set in the "academic" environment and therefore care must be taken not to overgeneralize the obtained results. Our conclusions need to be validated with professional programmers before they can be applied within the "industrial" environment. Yet university research of this type might very likely play a crucial role in the development of predictive models since it is impractical to expect industries to "repeat" a task.

Our research has concentrated on small programs (less than 500 lines of code) because of our environment. However it should be noted that over 60% of the "real world" programs fall into this category (Jones, 1981).

The obvious followup to the research presented in this study is an attempt to predict the parameters of the log-normal distribution based on some measurement of the assignment specification. The authors are currently investigating this topic with very promising preliminary results (McNicholl & Magel, in prep.).

APPENDIX

P.D.F.'s, C.D.F.'s, and Parameter Point Estimation Techniques for the Postulated Distributions

1. Normal Distribution

(a) Probability density function (p.d.f.)

$$f(x; \theta_1, \theta_2) = [\theta_2 \sqrt{2\pi}]^{-1} \exp[-.5((x - \theta_1)/\theta_2)^2],$$

(b) Cumulative distribution function (c.d.f.)

$$F(x; \theta_1, \theta_2) = [\sqrt{2\pi}]^{-1} \int_0^x (\theta_2)^{-1} \exp[-.5((x - \theta_1)/\theta_2)^2] \, dx$$

(c) Maximum likelihood parameter point estimators

$$\theta_1 = n^{-1} \sum_{i-1}^{n} X_i$$

$$\theta_2 = \left[(n - 1)^{-1} \sum_{i=1}^{n} (X_i - \theta_1)^2 \right]^{1/2}$$

2. Log-Normal Distribution

(a) Probability density function (p.d.f.)

$$f(x; \theta_1, \theta_2) = [x\theta_2\sqrt{2\pi}]^{-1} \exp[-.5((\ln(x) - \theta_1)/\theta_2)^2],$$

where $0 < x, \theta_2, -\infty < \theta_1 < +\infty$

(b) Cumulative distribution function (c.d.f.)

$$F(x; \theta_1, \theta_2) = [\theta_2\sqrt{2\pi}]^{-1} \int_0^x (x)^{-1} \exp[-.5((\ln(x) - \theta_1)/\theta_2)^2]\, dx$$

(c) Maximum likelihood parameter point estimators

$$\hat{\theta}_1 = n^{-1} \sum_{i=1}^n \ln(X_i)$$

$$\hat{\theta}_2 = \left[(n - 1)^{-1} \sum_{i=1}^n [\ln(x) - \theta_1]^2 \right]^{1/2}$$

3. Gamma Distribution

(a) Probability density function (p.d.f.)

$$f(x; \theta_1, \theta_2) = [\theta_1\Gamma(\theta_2)]^{-1} (x/\theta_1)^{\theta_2 - 1} \exp[-x/\theta_1]$$

where $0 < x, \theta_1, \theta_2$, and Γ is the gamma function.

(b) Cumulative distribution function (c.d.f.)

$$F(x; \theta_1, \theta_2) = [\Gamma(\theta_2)]^{-1} \int_0^x (x/\theta_1)^{\theta_2 - 1} \exp[-x/\theta_1](\theta_1)^{-1}\, Dx$$

(c) Maximum Likelihood parameter point estimators

$$\hat{\theta}_1 = \left[n^{-1} \sum_{i=1}^n X_i \right] \Big/ \hat{\theta}_2$$

The estimate for θ_2 is found by inverse interpolation of the following equation using the moment estimator of θ_2 as a starting value.

$$\ln(\theta_2) - \psi(\theta_2) - \ln\left[\left(n^{-1} \sum_{i=1}^n X_i \right) \Big/ \left(\prod_{i=1}^n X_i^{1/n} \right) \right] = 0$$

where the moment estimator for θ_2 is:

$$\theta_2 = \left[n^{-1} \sum_{i=1}^n X_i \right] \Big/ \left[(n - 1)^{-1} \sum_{i=1}^n \left(X_i - \left[n^{-1} \sum_{i=1}^n X_i \right] \right)^2 \right]$$

and ψ is the digamma function.

4. Beta Distribution

(a) Probability density function (p.d.f.)

$$f(x; \theta_1, \theta_2, \theta_3, \theta_4) = [B(\theta_3, \theta_4)]^{-1} \, (z)^{\theta_3 - 1} \, (1 - z)^{\theta_4 - 1} \, [\theta_2 - \theta_1]^{-1}$$

where

$$B(\theta_3, \theta_4) = \Gamma(\theta_3)\Gamma(\theta_4)/\Gamma(\theta_3 + \theta_4), \; z = (x - \theta_1)/(\theta_2 - \theta_1),$$
$$\theta_1 < x < \theta_2, \quad 0 < \theta_1 < \theta_2, \quad 0 < \theta_3, \theta_4$$

(b) Cumulative distribution function (c.d.f.)

$$F(x; \theta_1, \theta_2, \theta_3, \theta_4) = [B(\theta_3, \theta_4)]^{-1} \int_0^z (z)^{\theta_3 - 1} \, (1 - z)^{\theta_4 - 1} \, dz$$

(c) Maximum likelihood parameter point estimators

$$\hat{\theta}_1 = 0$$

It is especially difficult to develop an estimator for θ_2 since our process has no theoretic upper limit (see discussion in Section 5). For our purposes we will use the following ad hoc estimators:

$$\hat{\theta}_2 = \text{Max}(X_1, X_2, \ldots, X_n) + 1 \text{ if } X \text{ is discrete.}$$
$$\hat{\theta}_2 = \text{CEIL}[\text{Max}(X_1, X_2, \ldots, X_n)] \text{ if } X \text{ is continuous.}$$

The estimates for θ_3 and θ_4 must be found by an iterative procedure using the following two equations with the moment estimators as starting values.

$$\psi(\hat{\theta}_3) - \psi(\hat{\theta}_3 + \hat{\theta}_4) = \ln \left[\prod_{i=1}^{n} z_i^{1/n} \right]$$

$$\psi(\theta_4) - \psi(\theta_3 + \theta_4) = \ln \left[\prod_{i=1}^{n} (1 - z_i)^{1/n} \right]$$

where the moment estimators for θ_3 and θ_4 are:

$$\tilde{\theta}_3 = [(\bar{X})^2 - \bar{X} M_2]/[M_2 - (\bar{X})^2]$$
$$\tilde{\theta}_4 = [\bar{X} - M_2]/[M_2 - (M_2)^2] - \tilde{\theta}_3$$

and $\bar{X}$ is the sample average, and $M_2 = n^{-1} \sum_{i=1}^{n} (X_i)^2$

5. Type I Extreme Value Maxima Distribution

(a) Probability density function (p.d.f.)

$$f(x; \theta_1, \theta_2) = (\theta_2)^{-1} \exp[-z - \exp(-z)],$$

where

$$z = (x - \theta_1)/\theta_2, \quad 0 < x, \quad \theta_1 < \infty, \quad 0 < \theta_2$$

(*b*) Cumulative distribution function (c.d.f.)

$$F(x; \theta_1, \theta_2) = \exp[-\exp(-z)]$$

(*c*) Maximum likelihood parameter point estimators

$$\hat{\theta}_1 = -\hat{\theta}_2 \ln[n^{-1} \sum_{i=1}^{n} \exp(-x_i/\hat{\theta}_2)]$$

The estimate for θ_2 must be found by inverse interpolation of the following equation using the moment estimator as the starting value.

$$\hat{\theta}_2 - \bar{X} + \left[\sum_{i=1}^{n} x_i \exp(-x_i/\hat{\theta}_2) \right]\left[\sum_{i=1}^{n} \exp(-x_i/\hat{\theta}_2) \right]^{-1} = 0$$

where the moment estimator for θ_2 is:

$$\tilde{\theta}_2 = .7797\sigma_x$$

and σ_x is the standard deviation of the sample.

6. Weibull Distribution:

(*a*) Probability density function (p.d.f.)

$$f(x; \theta_1, \theta_2) = (\theta_2/\theta_1)(x/\theta_1)^{\theta_2-1} \exp[-(x/\theta_1)^{\theta_2}]$$

where:

$$0 < x, \theta_1, \theta_2$$

(*b*) Cumulative distribution function (c.d.f.)

$$F(x; \theta_1, \theta_2) = 1 - \exp[-(x/\theta_1)^{\theta_2}]$$

(*c*) Maximum likelihood parameter point estimators

$$\hat{\theta}_1 = \left[n^{-1} \sum_{i=1}^{n} (x_i^{\hat{\theta}_2}) \right]^{1/\hat{\theta}_2}$$

The estimate for θ_2 must be found by inverse interpolation of the following equation.

$$\left[\sum_{i=1}^{n} x_i^{\hat{\theta}_2} \ln(x_i) \right]\left[\sum_{i=1}^{n} x_i^{\hat{\theta}_2} \right]^{-1} - (\theta_2)^{-1} = n^{-1} \sum_{i=1}^{n} \ln(x_i)$$

REFERENCES

Bury, K. V. *Statistical models in applied science.* New York: John Wiley, 1975.
Chrysler, E. Some basic determinants of computer programming productivity. *Communications of the Association for Computing Machinery,* 1978, *21,* 472–482.

Curtis, B. *Human factors in software development.* IEEE Tutorial, COMPSAC November, 1981.

DeNelsky, G. Y., & McKee, M. G. Prediction of computer programmer training and job performance using the AABP test. *Personnel Psychology,* 1974, *27,* 129–137.

Faraguhar, J. A. A preliminary inquiry into the software estimation process. *Tech Report AD 712 052,* Defense Documentation Center, Alexandria, VA, August, 1970.

Jones, C. *Programming productivity: Issues for the eighties.* IEEE Tutorial, COMPSAC November, 1981.

Mann, N. R., Schafer, R. E., & Singpurwally, N. D. *Methods for statistical analysis of reliability and life data.* New York: Wiley, 1974.

McNamara, W. J., & Huges, J. L. A review of research on the selection of computer programmers. *Personnel Psychology,* 1961, *14,* 39–41.

McNicholl, D. G., & Magel, K. The Subjective Nature of Programming Complexity, presented at the Human Factors in Computer Systems conference, ACM, March, 1982.

McNicholl, D. G., & Magel, K. The Prediction of Resource Consumption during the Programming Phase of Software Development, in preparation.

Tonies, C. C. Project management fundamentals in software engineering. R. W. Jensen & C. C. Tonies (Eds.), *Software engineering.* Englewood Cliffs, NJ: Prentice-Hall, 1979.

Zelkowitz, M. V., Shaw, A. C., & Gannon, J. D. *Principles of software engineering and design.* Englewood Cliffs, NJ: Prentice Hall, 1979.

5

An Empirical Investigation of the Tacit Plan Knowledge in Programming*

KATE EHRLICH
ELLIOT SOLOWAY

Expert programmers know more than just the syntax and semantics of particular languages. This result has been demonstrated by a number of researchers who have replicated the classic chess experts of deGroot and Chase and Simon in the domain of programming. Oftentimes, however, experts aren't even aware of using this sort of knowledge, hence the term *tacit knowledge*. Building on this work, our goal is to identify *specific knowledge* which experts seem to have and use, and which novices have not yet acquired. In this chapter, we will describe the results of some initial exploratory studies using an experimental technique which seeks to tap into the specific tacit programming knowledge underlying variables.

We begin by highlighting our theory of programming knowledge underlying simple looping programs which we characterize in terms of stereotypic plans. In the experimental studies, we give programmers (novices and non-novices) a program in which important lines of code have been replaced by blank lines. Their task is to fill in the blank lines with appropriate lines of code. The results of the studies indicated that experienced programmers filled in the blanks correctly and in a plan-like fashion. We conclude by discussing some implications of this approach for teaching programming, building computer-based programming tutors, and developing cognitively based measures of program complexity.

What distinguishes a novice from an expert appears to be the fact than an expert possesses and uses *tacit knowledge* of the subject domain (Collins, 1978; Larkin, McDermott, Simon, & Simon, 1980; Polya, 1973). For example, an expert programmer would be able to see the underlying commonalities and differences among various problems and programs,

*This work was supported by the Army Research Institute for the Behavioral and Social Sciences, under ARI Grant No. MDA903-80-C-0508 and by the National Science Foundation, under NSF Grant SED-81-12403. Any opinions, findings, conclusions, or recommendations expressed in this report are those of the authors, and do not necessarily reflect the views of the U.S. Government.

or when to use specific programming constructions—knowledge which goes significantly beyond the syntax and semantics of particular programming languages. In this study, we will describe the results obtained from an empirical technique which seeks to elucidate the specific tacit knowledge differences between expert and novice programmers.

The organization of this study is as follows. We will first outline a fragment of a theory of programming knowledge, which serves as the theoretical basis for our experimental studies. We will then go on to describe the particular empirical technique that we have developed to examine programming knowledge. The technique is a new one to research on programming, but one that has correlates in research on text understanding. The studies, which fall into two groups, will then be presented. The first group of studies seeks to provide some validation of the technique, by examining whether the data we obtain are consistent with other research and general intuitions about programming. The second group of studies are more exploratory and are designed to address specific hypotheses about tacit programming knowledge. We conclude with some remarks about the implications of this type of research.

A THEORY OF PROGRAMMING KNOWLEDGE: PLANS

Evidence of the ability of experts to organize and structure knowledge has been demonstrated in a number of domains. For instance, building on the earlier work of de Groot (1965), Chase and Simon (1973) showed, in their classic study with Master and non-Master chess players, that the Masters could remember the board positions of pieces better than could the non-Masters when the board was arranged in some meaningful chess configuration. However, both Masters and non-Masters performed equally well when the configuration of the pieces was random. The authors attributed this result to the Masters' higher level knowledge about chess, which they used to chunk the board into easily remembered units. However, when the pieces were arranged at random, this knowledge was not as useful, and thus the Masters were no better off than the non-Masters. This experiment has been replicated in the domain of programming. Shneiderman (1976), Adelson (1981), and McKeithen, Reitman, Rueter, and Hirtle (1981), have shown that expert programmers can remember programs better than novices when the programs have some meaningful structure; but the experts do no better than the novices when the programs are composed of random lines of code. Again, the theory is that expert programmers used their higher level knowledge in order to code the presented programs for easier recall.

The studies on programming establish the point that experts seem to have more knowledge about programming and their knowledge is better organized than that of the novices. The studies, however, do not address the more important issue of the *content* of that knowledge. Our work at the Cognition and Programming Project at Yale has sought to identify the specific components of that underlying knowledge in order to better teach programming and in order to build computer-based environments which help novices to learn programming (Soloway, Woolf, Barth, & Rubin, 1981; Soloway, Rubin, Woolf, & Bonar, in prep.)

The basis of our approach is that expert programmers encode their higher level knowledge in the form of plans which represent many of the stereotypic actions in a program.[1] There are two main kinds of plan: CONTROL FLOW PLANS that represent looping structure and VARIABLE PLANS. For example, consider the Pascal program depicted in Figure 1, which computes the average of a set of numbers. This program uses a standard, common technique in programming: the *Running Total Loop Plan*.[2] This loop plan is used to accumulate partial totals and also keep track of the number of numbers. While expert programmers may not consciously call this structure the *Running Total Loop Plan,* we claim that programmers do, in fact, implicitly use this looping plan. In previous research we have examined the strategies required to implement looping plans in Pascal (Soloway, Bonar, & Ehrlich, 1983). However, here we are focusing on the kind of programming knowledge encoded in variable plans. We now turn to a discussion of these variable plans which will be used to motivate the experimental studies.

In addition to control flow plans, we suggest that programmers also make use, at least implicitly, of VARIABLE PLANS. A variable plan draws together various aspects surrounding the use of a variable.

- One aspect is the variable's *role* in the program, i.e., the function it serves. The program in Figure 1 includes examples of three different kinds of roles that variables can play. For example, the variable Count serves as a COUNTER VARIABLE, keeping track of the number of numbers read in. Sum serves as a RUNNING TOTAL VARIABLE, accumulating the sum of the numbers read in, while Number serves as a NEW VALUE VARIABLE, holding the new number read in each time through the loop.

---

[1]A similar approach has been adopted by other researchers who are interested in the structure of expert programming knowledge (e.g., Rich, 1981; Rich & Schrobe, 1978; Waters, 1979).

[2]Atwood and Ramsey (1978), based on their empirical studies, have also suggested that experts use this particular plan. In Soloway, Ehrlich, Bonar, & Greenspan (1983), we describe a network of such looping plans. (See also Rich, 1981; Waters, 1979; and Soloway & Woolf, 1979, for other sets of programming plans.)

Problem: Read in a set of integers and print out their average. Stop reading numbers when the number 99999 is seen. Do NOT include the 99999 in the average.

```
                        PROGRAM BlueAlpha;

                        VAR Sum, Count, Num : INTEGER;

                              Average : REAL;

Counter Variable       BEGIN
Plan          ----------> Count := 0;
              |    ---> Sum := 0;                    Running Total Loop Plan
              |    |    Read(Num); <--------------------
              |    |    WHILE Num <> 99999 DO <-------|
Running Total|     |    BEGIN                          |
Variable Plan|     |    ------> Sum := Sum + Num; <-------|
              |  -------------> Count := Count + 1;       |
              --------------->  Read(Num);        <--------
                        END                      Skip Guard Plan
                        IF Count > 0 THEN  <--------------------------
                              BEGIN <-----------------------|
                                 Average := Sum/Count; <-------|
                                 Writeln( Average); <----------|
                              END <-----------------------|
                              ELSE  <-----------------------|
                                 Writeln( 'no legal inputs'); <-|
              END
```

FIGURE 1. **A sample program: The running total loop plan.**

- A second aspect of a variable concerns the manner of its *initialization* and *update*. For example, both Count and Sum are initialized to 0, but the former is updated by 1 via an assignment, while the latter is updated by the value of the New Value Variable via an assignment; Number is initialized and updated via a Read.
- A third aspect is the *guard* which may protect a variable. For example, both Count and Sum need to be protected from including the sentinel value (99999) in their respective totals; the test in the "while" loop implements this guard.

These various aspects are intimately related to each other, and the plan structure serves to highlight this fact. For example, if a variable is *initialized* via an assignment (or Read), then it is typically *updated* via an assignment (or Read). Or, the *role* of a variable often suggests how it will be *initialized* and *updated;* e.g., by definition, a Counter Variable will start at some bottom value, and will keep track, via an assignment statement, of the number of entities of interest.

Information, such as the above, in a plan is not necessarily absolute; rather, plans usually reflect what is typical, and thus they can be overridden in specific situations. In particular, other elements in a program— the context—can also exert influences on such factors as the initialization and update of a variable. For example, in order to solve a problem, one might need to write a program in which a variable was initialized by a Read, but updated by an assignment. Being able to integrate contextual constraints with stereotypic plan knowledge is a sophisticated skill. We shall report on studies that explore the degree to which novices and non-novices are sensitive to contextual constraints.

In summary, then, a plan (control flow or variable) is a higher-level structure which serves to *chunk* together related pieces of information. Our notion of plan is akin to the notion of script, posited by Schank and Abelson (1977) to explain how people understand stereotypic sequences of events, such as going to the doctor or eating at a restaurant. Analogously, we believe that plan knowledge in programming consists of a catalog of stereotypic action sequences; Rich (1981) has even called such plans "cliches" to emphasize their stereotypic nature. On this view, then, programming, for an expert programmer, is an oftentimes complex process of selecting plans from his catalog and customizing them to the needs of the current situation. This catalog of plans is acquired through programming experience. For this reason, we expect that advanced programmers, who by definition have had more programming experience, should possess this sort of tacit plan knowledge to a greater degree than novices.

Goal of the Empirical Studies

The goal of the experimental studies is to examine whether there is empirical support for our theory of the tacit, plan knowledge underlying the use of variables. Based on the studies mentioned earlier, we would expect experts to have this kind of tacit knowledge better encoded than novices. Thus, support for our theory would be obtained if we found that expert programmers did perform better than novices, and if the former's responses were consistent with predictions made by our theory.

Our analysis of tacit plan knowledge will be used to guide the empirical studies. The first two studies will primarily focus on the features of variable plans that govern constraints between the role, the initialization, and the update of variables. Data from these studies should indicate the degree to which programmers respond to the more familiar constraints. The second and third studies will focus on sensitivity to the contextual constraints that are implicit in a program. Data from these two studies should indicate how well programmers respond to the more subtle contextual constraints that can serve to override constraints based on the more familiar relationships.

Methodology

In the studies reported here, we gave subjects an incomplete fragment of a program in which one or more critical lines had been left blank. Their task was to fill in the blank lines with one or more lines of code. We purposely did not tell the subject what problem the program was intended to solve. The point of the technique is to have the subject infer the intention of the program based on his or her knowledge of a typical problem and program plan. Hence, by carefully crafting both the program and which line(s) of code are left blank we can examine a subject's knowledge of a particular plan. Notice that if subjects had no understanding of the program, we would expect the answers they gave to be arbitrary, and thus we should see a great deal of variability. On the contrary, we find that there is a considerable degree of consistency in the answers that we get, particularly amongst the more experienced programmers.

A similar kind of fill-in-the-blank technique has recently been used in text comprehension research in order to tap into subjects' underlying knowledge of typical real world events. For example, Kemper (1982) presented subjects with a story in which gaps were created by deleting certain critical lines in the text and replacing them with blanks. She found that subjects used their knowledge of typical causal sequences to fill in the missing information. A similar result using an analogous technique has been obtained by Bower, Black, and Turner (1979). These results suggest that the fill-in-the-blank technique is appropriate to our goal of eliciting the kind of stereotypic knowledge that underlies programming, and that the technique can provide us with a window onto programmers' tacit knowledge.

A final word of caution is in order: doing empirical research on *tacit* knowledge is quite difficult because: (1) programmers are often unaware of using this sort of knowledge; (2) there is no universal agreement on what that knowledge is; and (3) one cannot examine the knowledge directly. Thus great care must be taken in identifying the specific knowledge that is being tested for, and in designing a task which actually taps into this knowledge, and hence to go beyond restating the obvious; "that experts are experts because they know more."

THE EXPERIMENTAL STUDIES

The four studies that we will describe below were part of a larger test that included other programming problems in Pascal. The test was administered to three different groups of programmers. There was a group of 96 novice programmers who were at the end of their first semester course in programming in Pascal; a group of 35 intermediates who were at the end of their second semester course in programming; and 18 advanced programmers who were at the end of their third to fifth course in programming. Within the larger test each subject received four fill-in-the-blank problems which were randomly interspersed with the other programming problems in the larger test. The whole test took about an hour to complete. The task given to the subjects was to fill in the blank lines with one or more lines of code that would, in their opinion, "best complete" the program. Subjects were given the option of not filling in a line if they believed that there was nothing that should go there.[3] They also had the option of filling in more than one line of code for each blank line, an option that a few subjects did take up. In some of the problems there were two lines left blank and in others there was only one blank line.

The names given to the fill-in-the-blank problems were Green, Red, Blue, and Violet. There is no significance attached to these names, except to eliminate any clue as to the program's function. We will use these names to simplify reference to the different problems. Problems Green and Red were designed to explore the degree to which the role, initialization, and update of a variable influences the way a programmer fills in the blank lines in the program, by examining responses to a variety of different kinds of variable plans. Problems Blue and Violet, in contrast, were designed to explore specific hypotheses concerning the degree to which contextual constraints interact with plans.

---

[3]In the data we will report we have only included the responses of those subjects who filled in at least one of the blanks. We have excluded those subjects who left two lines blank since we have no way of knowing whether they failed to do the problem because it was too hard, because they were bored, because they accidentally turned over two pages of the booklet and failed to see the problem at all, or any number of other reasons that do not directly reflect either their competency or the problem's difficulty. For this reason, the total number of subjects in each group may vary slightly between the studies.

PROBLEMS GREEN AND RED: THE INITIALIZATION OF VARIABLES

A further goal behind the first two studies was to evaluate the fill-in-the-blank technique on simple problems. Before we can use the technique to examine issues concerning programming knowledge, we need to know that the technique is sensitive to some of the more obvious aspects of programming behavior. Initialization of variables is a basic concept in programming, and one that should not give people much difficulty. However, novices should have more difficulty than experienced programmers.

Problem Green: Initialization of Counter Variables

Problem Green, Figure 2, is a program that adds up the number of characters and the number of blanks in some input line. The number of characters is accumulated in a Counter Variable called SymbolCnt, and the number of blanks is accumulated in another Counter Variable called BlankCnt.

A perusal of the program reveals that neither of these variables has been initialized. Since the program is to add up the number of entities

Please complete the program fragment given below by filling in the blank line (indicated by a box). Fill in the blank with a line of Pascal code which in your opinion best completes the program.

```
program Green (Input/, Output);
const Blank = ' ';
var SymbolCnt, BlankCnt : integer;
    Symbol : char;
begin
---------------------------------
|                               |;
---------------------------------
---------------------------------
|                               |;
---------------------------------
while not eoln do
  begin
  Read(Symbol);
  if Symbol = Blank
    then BlankCnt := BlankCnt + 1;
  SymbolCnt := SymbolCnt + 1
  end;
Writeln (SymbolCnt, BlankCnt);
end.
```

FIGURE 2. **Problem Green: Initialization of a Counter Variable.**

TABLE 1
Initialization of Counter Variables: The Number of Novices and
Non-novices Who Correctly Initialized Both Counter Variables[a]

	Novices	Non-novices
Correct (initialized BlankCnt and SymbolCnt to 0)	61	50
Incorrect	31	3

[a]$\chi^2 = 14.72$, $p < 0.001$.

meeting a specified property, and since the Counters are updated after the entity has been found, the Counters should be initialized to 0; if they were initialized to 1, their final value would be "off by one." Table 1 shows the number of subjects who correctly initialized the counter variables. Since there was no difference in performance between the intermediates and advanced groups, we have combined their results into a non-novice category. Although most of the novices (66%) did initialize both variables correctly, they did so significantly less often than the more experienced (non-novice) subjects (94%) ($\chi^2 = 14.72$, $p < .001$). The most common alternative used by the novices was to fill in the blanks with some combination of Read and Write lines. In contrast, our data suggest that non-novice programmers have no difficulty in recognizing that the Counter Variables need to be initialized via an assignment statement (as opposed to, say, via a Read). Thus, they performed as if they did understand the relationship between initializing and updating Counter Variables. The data also suggest that most, but by no means all, of the novices also understand this relationship. In the next section, we will explore their understanding of how Running Total Variables and Limit Variables should be initialized.

Problem Red: The Limit and Running Total Variables

The objective of the program in Problem Red (Figure 2) is to read a set of numbers and compute their average. Here, both N, the number of numbers, and Sum, the Running Total Variable, need to be initialized. Table 2 shows the number of people in each group who correctly initialized the Running Total Variable (e.g., Sum := 0). The figures are almost identical to the ones we obtained for the Counter Variable in Problem Green. Again, the novices perform poorer than the more experienced programmers. The novices filled in the blank correctly only 70% of the time whereas the non-novices were correct 92% of the time ($\chi^2 = 10.86$, $p < .01$).

Please complete the program fragment given below by filling in the blank line (indicated by a box). Fill in the blank with a line of Pascal code which in your opinion best completes the program.

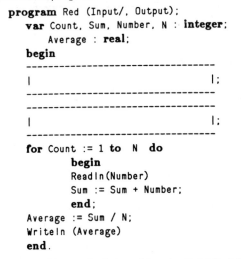

```
program Red (Input/, Output);
    var Count, Sum, Number, N : integer;
        Average : real;
    begin
    ----------------------------------
    |                                I;
    ----------------------------------
    ----------------------------------
    |                                I;
    ----------------------------------
    for Count := 1 to  N  do
            begin
            ReadIn(Number)
            Sum := Sum + Number;
            end;
    Average := Sum / N;
    Writeln (Average)
    end.
```

FIGURE 3. **Problem Red: Limit and Running Total Variables.**

In contrast, the Limit Variable N is not updated via an assignment, but rather it is used to control the looping in the **for** statement. How then ought it to be initialized? We contend that N should be initialized via a Read rather than via an assignment (e.g., $N := 10$) based on the following argument:

> If N were to be set via an assignment statement, e.g. $N := 10$, then N would be used as a constant, and if so, then either the actual, constant value should be substituted into the for loop (replacing N), or N should be declared as a constant (CONST) in the declaration section of the program.

This is a rather sophisticated argument which we felt that only the most experienced programmers would appreciate and have internalized.

TABLE 2
Initialization of Running Total Variables: The Number of Novices and Non-novices Who Correctly Initialized the Running Total Variable[a]

	Novices	Non-novices
Correct (initialized Sum to 0)	62	47
Incorrect	27	4

[a]$\chi^2 = 10.86$, $p < 0.01$.

TABLE 3
Setting N via a Read or a Constant: Categorization of Responses by Group

	Novice	Intermediate	Advanced[a]		Novices	Non-novices[b]
Read N	32	14	16	Correct	45	46
Set N to a Constant	13	15	1			
Nothing done to N	44	4	1	Incorrect	44	5

[a] $\chi^2 = 37.55$, $p < .001$.
[b] $\chi^2 = 22.39$, $p < .001$.

Unlike the case of initializing the Counter Variable or the Running Total Variable, there is no one right answer for how to initialize the Limit Variable: while setting N via a Read is the answer we argue is best, setting N via an assignment statement is certainly not wrong. Note further that the above argument for setting N via a Read is based on a notion of *communication:* the writer of a program intends to convey more than just the literal meaning of the program. In particular, by using a variable as the stopping value of the **for** loop, the programmer is saying *"the stop value will change from time to time in this program."* This communication information is part of the tacit knowledge which expert programmers share. We will see this implicit communication again in Problem Violet.

The distribution of responses across the three groups is shown in Table 3.[4] The data indicate that the dominant response amongst the novices was to fail to set N at all. An analysis comparing correct responses (either head N or setting N to a constant) indicated that the novices were correct only 51% of the time whereas the non-novices were correct 90% of the time ($\chi^2 = 22.39$, $p < .001$). Among the intermediates, a bit more than half set N via a constant. Among the advanced programmers, the dominant response was to read in the value of N. This difference between the three groups in their category of response was significant ($\chi^2 = 37.55$, $p < .001$). Hence, the data are consistent with the argument, put forth above, that advanced programmers do perceive that a Limit Variable should be initialized via a Read.

Summary: Problems Red and Green

The data from problems Green and Red indicated that, as we expected, programmers with at least 2 semesters experience of programming gave responses that were consistent with them understanding the constraints

[4]We separated the intermediate and advanced groups to show which correct response each group selected.

that operate between the role, initialization, and update of variables. The responses of the novices were much more erratic than the other programmers. This kind of variability would be expected from a group that lacked the underlying tacit, plan knowledge.

PROBLEMS BLUE AND VIOLET: CONTEXT EFFECTS

In addition to understanding the influence of a variable's role, and update on its initialization, the program context can also exert an important influence on the manner in which a variable is initialized or updated. Reacting appropriately to contextual constraints is a sophisticated ability. Thus, we hypothesize that novices, with their limited experience, will be less sensitive to such constraints than would non-novices. In the next two studies we will report some initial experiments which begin to explore this hypothesis.

PROBLEM VIOLET: RESOLVING CONFLICTS BETWEEN PLANS

In this study, we explore the degree to which programmers use contextual constraints to resolve a conflict between two competing plans in one program. Our hypothesis is that novices will not be as sensitive to the contextual constraints as would nonnovices, but will rather fall back on a plan with which they are more familiar. In contrast, nonnovices should be able to avoid being seduced by a more familiar plan; rather, they will have the ability to use context to resolve any conflicts.

The programs in Figures 4 and 5 are both intended to produce the square root of N, the New Value Variable. Since N is in a loop which will repeat 10 times, 10 values will be printed out. The question is: How should N be set? The technique will be to compare the performance of programmers on the program which does not contain the plan conflict with their performance on the program which does contain the conflict.

Consider first the program in Figure 4, which is the version that does not contain the intended plan conflict. N is a New Value Variable, since its function is merely to hold successive values. The plan for this type of variable does not present an overriding constraint on how it should be set in the blank line: a Read(N) or a N := N + SomeValue would both be acceptable.

However, the context does provide a strong constraint. Notice that the **if** test preceding the Sqrt(N) instantiates the *"guard a portion of a program from improper data"* plan by protecting the Sqrt from negative integers (the Sqrt function can only work on positive integers). This test specifies an important constraint: N should take on values that could possibly be negative, otherwise the **if** test would be totally superfluous. Thus, N should not be set via an assignment statement to some simple

Please complete the program fragment given below by filling in the blank line (indicated by a box). Fill in the blank with a line of Pascal code which in your opinion best completes the program.

```
program VioletAlpha(Input/, Output);
  var N :real;
      I : integer;
  begin
  for I := 1 to 10 do
    begin
    ---------------------------------
    |                               |;
    ---------------------------------
    if N < 0 then N := -N;
    Writeln ( Sqrt(N) );
       (* Sqrt is a built-in
          function which returns the
          square root of its argument*)
    end;
  end.
```

FIGURE 4. **Problem VioletAlpha: The influence of a single plan.**

Please complete the program fragment given below by filling in the blank line (indicated by a box). Fill in the blank with a line of Pascal code which in your opinion best completes the program.

```
program VioletBeta(Input/, Output);
  var N :real;
      I : integer;
  begin
  N := 0.0;
  for I := 1 to 10 do
    begin
    --------------------------------
    |                              |;
    --------------------------------
    if N < 0 then N := -N;
    Writeln ( Sqrt(N) );
       (* Sqrt is a built-in
          function which returns the
          square root of its argument*)
    end;
  end.
```

FIGURE 5. **Problem VioletBeta: A conflict between plans.**

function of N and/or the index variable I, e.g., $N:=N+I$, $N:=I$, $N:=N+1$. Rather, by setting N via a Read statement, negative values have the possibility of entering the program. Note that this argument is, like the argument why Read was the most appropriate method for setting the

TABLE 4
Conflict between Plans: Assignment (familiar) vs. Read (guarding)

	Novice[a]		Non-novice[b]	
Category	ALPHA (no conflict)	BETA (conflict)	ALPHA (no conflict)	BETA (conflict)
1. Set N via Read	44	30	20	26
2. Set N via Assignment	7	15	4	4

[a]$\chi^2 = 5.20$, $p < 0.05$.
[b]$\chi^2 < 1$, N.S.

Limit Variable in Study Red, one based on a principle of tacit communication:

> by including a test for negative values, an experienced programmer is informing the reader that it is possible that such numbers could be generated; if such numbers could not possibly enter the program, then the inclusion of this test would violate an unwritten, but yet well understood rule of communication.

The blank line in the program in Figure 4 is strategically placed: we wanted to explore the degree to which programmers are sensitive to the contextual relationship which obtains between the guard plan and the initialization aspect of New Value Variable Plan.

Program VioletBeta in Figure 5 is exactly the same as that in Figure 4, except that now N is given a value of 0 before the loop. Previously the New Value Variable Plan was neutral with respect to how N should be set. However, since N was initialized via an assignment statement to 0, the general rule of relating initialization to update should come into play, and direct that N be updated via an assignment. On the other hand, the **if** test, which realizes the "guard plan" and protects the square root operation, still sets up the expectation that N will be read in. If N will be set via a Read in the loop, the setting of N to 0 initially is superfluous. Thus, in Program VioletBeta we have purposely created a situation in which two plans are in conflict: the New Value Variable Plan expects N to be updated via an assignment, since it was initialized via an assignment, but the guard plan on the Sqrt operation expects that N will updated via a Read, so as to permit negative values to enter the program.

The data we obtained from problems Green and Red indicated that novices do seem to recognize the constraint between the initialization and update aspects of a variable. Given our understanding of the kinds of programs to which they tend to be exposed, we felt that novices would also have more experience with this plan than with the guard plan. On the other hand, we felt that more advanced programmers would have had sufficient experience in both, and know when each is most appropriate. Thus, we hypothesized that non-novice programmers would realize that the test for a negative N would take precedence over the initialization of N to 0, since the "guarding" of the input is usually very important to the correct running of the program. Thus, we predicted that nonnovices would fill in the blank with Read(N) equally often in both versions of the problem. In contrast, we felt that novices, with their limited experience, would not be as sensitive as non-novices to the importance of the test for a negative N. Hence, we predicted that the proportion of novices who Read in the value of N would decline when there was a conflict between plans.

Since we were interested in the manner in which the variable N was set in the loop, we constructed two categories for scoring the responses. Category 1 indicated that N was set via a Read, while category 2 indicated that N was set via an assignment statement. Except for one case, we grouped together all forms of assignment. Two novices filled the blank in Program Beta with $N := N - I$. This is a clever answer since it satisfies both constraints: N is updated by an assignment in agreement with the initialization, and N can take on negative values in agreement with the *if* test. Since these two answers satisfied both constraints, unlike the answers in categories 1 and 2, we did not include these two answers in either of the categories. In principle any answer that allows N to take on negative values would satisfy both constraints, however, the response $N := N - 1$ was the only example of this type that we found.

The data, shown in Table 4, support our predictions. Non-novices chose to set N via a Read in the no conflict case (Alpha), and also chose to set N via a Read in the conflict case (Beta). This is consistent with our hypothesis that non-novices could use contextual information—the guard plan constraint—to override the variable plan constraint in the conflict case. In contrast, novices chose Read significantly less often in the conflict case than in the nonconflict case ($\chi^2 = 5.20$, $p < .05$).[5] This is consistent with our hypothesis that novices were more influenced by the familiar variable plan constraint than by the less familiar, contextual guard plan constraint.

Problem Blue: Conflicting Looping Plans

It as been observed (e.g., Knuth, 1974; Wegner, 1979) that Pascal's **while** loop often requires an awkward coding strategy. A typical case is a

[5]We have also run this study with a group of 87 introductory BASIC students, and obtained the same pattern of data. That is, the students filled in the blank with an assignment statement more frequently when there was a conflict than when there was no conflict ($\chi^2 = 14.64$, $p < .001$).

program which repeatedly reads in integers until some integer, say 99999, is read in, and then print out the average of the numbers. An example of a correct program is shown in Figure 1. In this program, a value is read into the New Value Variable, Number, outside the loop and tested by the **while** construct before the loop is executed even once, since the sentinel value 99999 can appear on the first read. Within the loop the sequence of operations is: add the number to the total, increment the counter, and only then, read in the next number.

We have called the looping strategy which underlies this implementation a "process i/read next-i" strategy: on the ith pass through the loop, the ith value is processed and the next-ith value is read in. In other words, processing in the loop is out of synchronization with the read. We have shown that programmers of varying levels of experience do not naturally employ a process i/read next-i strategy, but rather, they do employ a read i/ process i strategy: on the ith pass through the loop, the ith value is read and processed (Soloway, Bonar, & Ehrlich, 1983). However, if one wants to master Pascal, one needs to internalize the plans for using the **while** construct, regardless of its proven conceptual awkwardness.

In this study we will be contrasting knowledge of the correct strategy with knowledge that derives from nonprogramming experience. An important aspect of the tacit knowledge of variables is the ability to use this knowledge to override simpler strategies. In problem Violet, we saw that experienced programmers were sensitive to the need to satisfy the guard plan in preference to maintaining consistency in the manner in which a variable was updated. In this problem, we are similarly interested in whether experienced programmers are sensitive to the context provided by the looping construct, by knowing where in the program the input variable should be mentioned. We expect that novices will be less adept than non-novices at overriding their natural tendency toward filling in the blank with a response that is consistent with a read/process strategy.

We will examine these claims by contrasting two similar programs. Consider first the program in Figure 6, in which there are two blank lines. One must be used for the Read on the New Value Variable, and the other can reasonably be left empty. We felt that the heart of the **while** process i/read next-i plan was given in this program and that all our programmers would be able to recognize the need to fill in one of the blank lines with Read(Number). In contrast, notice that one of the blank lines in the program in Figure 7 is inside the loop. Quite reasonably, again, this line could be left blank. Our hypothesis, however, was that novices would revert to the more natural read i/process i strategy and fill the blank inside the loop with Read(Number). However, the

Please complete the program fragment given below by filling in the blank line (indicated by a box). Fill in the blank with a line of Pascal code which in your opinion best completes the program.

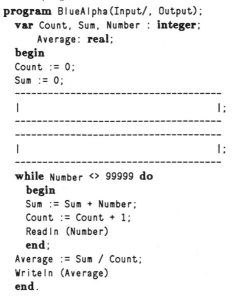

```
program BlueAlpha(Input/, Output);
   var Count, Sum, Number : integer;
      Average: real;
begin
Count := 0;
Sum := 0;
------------------------------------
|                                  |;
------------------------------------
------------------------------------
|                                  |;
------------------------------------
while Number <> 99999 do
  begin
  Sum := Sum + Number;
  Count := Count + 1;
  Readln (Number)
  end;
Average := Sum / Count;
Writeln (Average)
end.
```

FIGURE 6. **Problem BlueAlpha: All blanks outside the loop.**

non-novices would have the experience and knowledge to know to override that response, and rather to fill the blank outside the loop with Read(Number).

The results in Tables 5 and 6 bear out these predictions. In Table 5 we see that novices do as well as non-novices on the program in Figure 6, the program in which both blank lines are outside the loop. However, notice that on the program in Figure 7 the novices tend to fill in the blank *inside* the loop where non-novices still, correctly, fill in the blank outside the loop with a Read statement. That is, the novices did significantly worse on the program in Figure 7 than on the program in Figure 6. In Table 6 we break the results down further by categorizing the type of error which was committed. In the case of the novices, we see that they were often inclined to put the Read on the New Value Variable inside the loop, whereas the non-novices were not so inclined.[6]

---

[6]Interestingly, of the 33 novices and the 25 intermediate and advanced people who did read in a number only 15% of the novices left the other line blank as compared with 52% of the other programmers ($\chi^2 = 9.02$, $p < .02$). Thus, the more advanced programmers seem to know that there is nothing more that needs to be added. The novices are less confident about leaving the line blank.

TABLE 5
Errors in Loops: Accuracy of Response for Novices and Non-novices

	Novice[a]		Non-novice[b]	
Category	Alpha (both blanks outside)	Beta (one blank in, one blank out)	Alpha (both blanks outside)	Beta (one blank in, one blank out)
1. Correct (Read statement outside loop)	33	28	25	19
2: Wrong	7	25	5	3

[a]$\chi^2 = 8.89$, $p < .01$.
[b]$\chi^2 < 1$, N.S.

Please complete the program fragment given below by filling in the blank line (indicated by a box). Fill in the blank with a line of Pascal code which in your opinion best completes the program.

```
program BlueBeta(Input/, Output);
    var Count, Sum, Number : integer;
        Average: real;
    begin
    Count := 0;
    Sum := 0;
    ------------------------------------
    |                                  |;
    ------------------------------------
    while Number <> 99999 do
        begin
        ------------------------------------
        |                                  |;
        ------------------------------------
        Sum := Sum + Number;
        Count := Count + 1;
        Readln (Number)
        end;
    Average := Sum / Count;
    Writeln (Average)
    end.
```

FIGURE 7. **Problem BlueBeta: Choice between correct and natural strategy.**

Summary: Problems Violet and Blue

The results of these two studies serve to demonstrate the differences between novice and non-novice programmers in their sensitivity to the contextual constraints that are implicit in a program. In both studies we

TABLE 6
Errors in Loops: The Responses of Novices and Nonnovices as a Function of Whether
There was a Blank Inside the Loop (BETA) or Not (ALPHA)

	Novice		Nonnovice	
Category	Alpha	Beta	Alpha	Beta
1: Correct (Read statement outside loop)	33	28	25	19
2: Wrong (Read statement inside loop)	N/A	16	N/A	1
3: Wrong (No Read statement)	7	9	5	2

saw that the experienced programmers filled in the blanks in a way that
was consistent with the more general, and often more subtle kind of
plan. The novices, however, often seemed to fill in the blanks with code
that was consistent with more familiar aspects of variables. These data
provide further evidence that plan based knowledge encodes the kind of
contextual constraints that can be used to override more familiar
strategies.

CONCLUDING REMARKS

The experimental studies were designed to provide an exploratory empirical examination of the tacit, plan knowledge of variables which we
outlined in Section 3. In all four studies, the responses of the advanced
programmers[7] were consistent with predictions based on our theory of
tacit, plan knowledge. This novice/expert performance difference is expected, since the knowledge we claim that experts are using is *tacit*
knowledge which is typically acquired through experience. We are, however, well aware that the data we obtained are open to alternative interpretations. Nevertheless, it must be said that the data are consistent
with the theory and that we have tried to specify the theory in detail
sufficient for testing by independent researchers. More research is
needed in this field before we can fully understand and explicate the
knowledge that underlies programming. Indeed, we would hope that

---

[7]The programmers in this study were students at a major university, not professional
programmers. While some of our subjects in the advance group had extensive programming experience, care must be taken in generalizing the results reported here.

our studies might provoke other researchers to carry out further experiments on this topic.

The research in this study represents a shift in focus from other recent empirical work on programming (Adelson, 1981; McKeithen et al., 1981; Shneiderman, 1976). Other approaches have often sought just to demonstrate that experts have more and better organized knowledge than do novices. In contrast, our approach is to identify the specific knowledge which expert programmers have (see also Rich, 1981; Soloway, Ehrlich, Bonar, & Greenspan, 1983).

The product of our approach can directly benefit a number of areas in programming:

- *Teaching programming:* Rather than emphasizing the syntax and semantics of a particular programming language, we suggest that these plan notions should be incorporated into texts and classroom instruction (e.g., see Soloway & Woolf, 1979). Also, by looking at the knowledge which is needed for a correct solution, we have a better understanding of the misconceptions which students develop (Soloway, Bonar, Woolf, Barth, Rubin, & Ehrlich, 1981; Soloway, Ehrlich, Bonar, & Greenspan, 1983).

- *Building computer-based programming tutors:* We are building a family of computer-based tutors which can find nonsyntactic bugs in students' programs, and provide students with feedback on their underlying misconceptions (Soloway, Woolf, Barth, & Rubin, 1981; Soloway, Rubin, Woolf, & Bonar, in prep.). The basis for these systems is the theory of plan knowledge we are developing which provides a rich vocabulary in which to analyze programs, diagnose misconceptions, and provide useful information to the student.

- *Developing cognitively-based measures of program complexity:* One of our longer range goals is the development of measures of program complexity based not just on features of the program text itself, but rather on the cognitive demands which the program makes on the programmer. Sebrechts and Black (1982) have argued quite persuasively that measures of program complexity based on textual features (e.g., number of operations, length of variable names) cannot be effective measures, in the same way that the old measures of reading complexity, based also on textual features, were not effective measures. Such measures can capture only "surface" information. In contrast, effective measures must be based on the types and number of inferences which a programmer must make in order to understand the program. First, however, we need to catalog the knowledge which a programmer uses to understand a program—the goal of this research.

REFERENCES

Adelson, B. Problem solving and the development of abstract categories in programming languages. *Memory and Cognition*, 1981, *9*, 422–433.

Atwood, M. E., & Ramsey, H. R. Cognitive structure in the comprehension and memory of computer programs: An investigation of computer program debugging. Tech. Rept. ARI TR-78-A210, Science Applications, Englewood, CA, 1978.

Bower, G. H., Black, J. B., & Turner, T. Scripts in memory for text. *Cognitive Psychology*, 1979, *11*, 177–220.

Chase, W. C., & Simon, H. Perception in chess. *Cognitive Psychology*, 1973, *4*, 55–81.

Collins, A. Explicating the tacit knowledge in teaching and learning. Tech. Rept. 3889, Bolt Beranek and Newman. Cambridge, MA, 1978.

deGroot, A. D. *Thought and choice in chess*. Paris: Mouton, 1965.

Kemper, S. Filling in the missing links. *Journal of Verbal Learning and Verbal Behavior*, 1982, *21*, 99–107.

Knuth, D. Structured programming with GO TO statements. *ACM Computing Surveys*, 1974, *6*, 4.

Larkin, J., McDermott, J., Simon, D., & Simon, H. Expert and novice performance in solving physics problems. *Science*, 1980, *208*, 140–156.

McKeithen, K. B., Reitman, J. S., Rueter, H. H., & Hirtle, S. C. Knowledge organization and skill differences in computer programmers. *Cognitive Psychology*, 1981, *13*, 307–325.

Polya, G. *How to solve it*. Princeton, NJ: Princeton University Press, 1973.

Rich, C. Inspection methods in programming. Tech. Rept. AI-TR-604, MIT AI Lab, 1981.

Rich, C., & Shrobe, H. Initial report on a LISP programmer's apprentice. *IEEE Transactions on Software Engineering*, 1978, *4*, 342–376.

Schank, R. C., & Abelson, R. *Scripts, plans, goals, and understanding*. Hillsdale, NJ: Erlbaum, 1977.

Sebrechts, M. M., & Black, J. B. Software psychology: A rich new domain for applied psychology. *Applied Psycholinguistics*, 1982, *3*, 223–232.

Shneiderman, B. Exploratory experiments in programmer behavior. *International Journal of Computer and Information Sciences*, 1976, *5*, 123–143.

Soloway, E., & Woolf, B. Problems, plans, and programs. Proceedings of Eleventh ACM Technical Symposium on Computer Science Education, ACM, 1979.

Soloway, E., Woolf, B., Barth, P., & Rubin, E. MENO-II: Catching run-time errors in novice's Pascal programs. Proceedings of IJCAI-81, International Joint Conference on Artificial Intelligence, Vancouver, BC, 1981.

Soloway, E., Bonar, J., Woolf, B., Barth, P., Rubin, E., & Ehrlich, K. Cognition and programming: Why your Students write those crazy programs. Proceedings of the National Educational Computing Conference, NECC, No. Denton, TX, 1981.

Soloway, E., Ehrlich, K., Bonar, J., & Greenspan, J. What do novices know about programming? In B. Shneiderman & A. Badre (Eds.) *Directions in human-computer interactions*, Norwood, NJ: Ablex, 1983.

Soloway, E., Bonar, J., & Ehrlich, K. Cognitive strategies and looping constructs: An empirical study. *Communications of the ACM*, 1983, *26*, 853–860.

Soloway, E., Rubin, E., Woolf, B., & Bonar, J. MENO-II: An intelligent programming tutor. *Journal of Computer-Based Instruction*, in prep.

Waters, R. C. A method for analyzing loop programs. *IEEE Trans. on Software Engineering* SE-5 May 1979, 237–247.

Wegner, P. Programming languages—Concepts and research directions. In *Research directions in software technology*. Cambridge, MA: MIT Press, 1979.

6

An Empirical Evaluation
of Software Documentation Formats

SYLVIA B. SHEPPARD
JOHN W. BAILEY
ELIZABETH KRUESI BAILEY

Documentation of program specifications can be categorized along two dimensions, type of symbology and spatial arrangement. Four types of symbology (normal English, ideograms, a program design language (PDL), and abbreviated English) were factorially combined with three spatial arrangements (sequential, branching, and hierarchical) to produce 12 documentation formats. Three experiments determined the optimal formats for coding, debugging, and modification tasks, respectively.

Substantial differences in performance appeared to be related to the succinctness of the various forms of symbology: the more succinct the symbology, the better the performance. In particular, normal English, the most verbose symbology, was associated with longer performance times than the more succinct PDL.

The effect of spatial arrangement was much less pronounced. However, the branching spatial arrangement was mildly superior to the other arrangements, particularly in reducing the number of errors related to control flow.

The success of any software development project depends in part on the quality of the communication among the individuals involved: users, systems analysts, designers, coders, and managers. This is a particularly critical factor in the development of a large system since a variety of individuals perform various related tasks at different times. The efficiency and quality with which later tasks are performed depends critically on the documentation supplied during previous phases of the development cycle. Thus, both managers and programmers alike are interested in the relative merits of the many types of documentation currently in use. Included among these are English prose, flowcharts, and program design languages (PDLs).

English prose is the most commonly used method for expressing problem specifications. However, an experiment recently reported by Miller (1981) raises some doubts about the advisability of normal English as either a design or documentation tool. Miller asked nonprogrammers to write normal English procedures for solving problems that were representative of common computer applications. Careful analysis of the protocols led Miller to conclude that even minor increases in the complexity of problems led to marked decreases in the quality of the solutions. Further, the high degree of contextual referencing found in the solutions provided doubts about the feasibility of adequate English specifications.

Flowcharts have often been used to structure problems and have been described as "an essential tool in problem-solving" (Bohl, 1971, p. 53). However, the empirical evidence concerning the usefulness of flowcharts is conflicting. An early empirical assessment of the value of flowcharts in programming was reported by Shneiderman, Mayer, McKay, and Heller (1977). They performed a series of experiments on the composition, comprehension, debugging and modification of programs. These experiments were conducted with student programmers who worked on modular-sized programs. For the composition task, the participants were asked to write a program. Some were asked to produce a flowchart in addition to the program. For the comprehension, debugging, and modification tasks, all participants were given a program listing, while some were given a flowchart as an additional aid. Shneiderman et al. found no significant differences in any of their experiments between groups that did and did not use flowcharts.

In another study, Ramsey, Atwood and Van Doren (1978) compared the effectiveness of flowcharts to that of a program design language. Ramsey et al. found no difference in the ease with which these two types of documentation could be comprehended. They did, however, find an advantage for PDL as a design tool.

Brooke and Duncan (1980) compared flowcharts and sequential English instructions as debugging tools. They concluded that flowcharts were useful for tracing execution sequences in a program but were not helpful in conceptualizing relationships among noncontiguous segments of a program.

Although studies performed on software-related tasks have not been especially favorable to flowcharts, experiments performed in other areas of information presentation have demonstrated an advantage for flowcharts over alternative presentation formats including prose descriptions, short sentences and decision tables. Wright and Reid (1973) presented participants with a set of hypothetical problems. The task was to choose from among several means of space travel. The correct answer was contingent upon cost, time, and distance factors. These contingencies were expressed in the form of a flowchart, a decision table, short sentences, or a bureaucratic-style (purposely verbose) prose. On easy problems, the participants were equally accurate with the flowchart, decision table, and short sentences; all three forms were superior to the bureaucratic-style prose. On more difficult problems, the participants were most accurate with the flowchart, intermediate with the decision table and least accurate with the sentences and bureaucratic prose.

Similar results were obtained by Blaiwes (1974), who presented participants with instructions for using a communications control console. These instructions were arranged in either a flowchart or a short sentence format. Blaiwes found no difference among these formats for easy problems. However, on difficult problems, use of the flowchart resulted in fewer errors.

Kammann (1975) presented a group of housewives and a group of scientists with a set of telephone-dialing problems. The dialing instructions were presented in the form of a prose description or in a flowchart. Both groups made fewer errors with the flowchart.

Some of the experiments cited here suggest that complex procedural information is presented more effectively in a lexical–graphical format than in a purely lexical format. Other studies indicate no differences. Given these discrepancies, we felt that a reexamination of the factors comprising various formats was in order. In particular, our interest was in evaluating how various characteristics of presentation affect the performance of programmers on typical software tasks.

We selected two primary dimensions for categorizing how available documentation aids configure the information they present to programmers. The first dimension is the type of symbology in which information is presented. The second dimension is the spatial arrangement of this information. Our approach to evaluating various forms of documentation was to investigate the separate and combined effects of the type of symbology and the spatial arrangement. The interrelation of these two dimensions describes both familiar and novel types of documentation. By expanding our realm of study beyond a comparison of only familiar formats, we hoped to discover more general principles which will aid software developers in selecting from the many available documentation aids as well as guide in the development of new aids.

TYPE OF SYMBOLOGY

The symbologies we investigated included ideograms, normal English, a program design language (PDL), and an abbreviated English. Ideograms are graphic symbols frequently found in flowcharts and hierarchical input–process–output (HIPO) charts (Bohl, 1971; Katzen, 1976). A standard set of ideograms has come to represent processes or entities within a program. Normal English is frequently used for documentation of all kinds. Program design languages are more succinct than normal English. The PDL in these experiments used variable names, symbols, and keywords to describe the procedural logic of the programs. The abbreviated English was identical to normal English, with the exception that the variable names were used rather than normal English descriptions of the variables. Thus, the abbreviated English was more succinct than the natural language but less succinct than the PDL.

SPATIAL ARRANGEMENT

The spatial arrangement of information in documentation is a second dimension along which documentation techniques can be categorized. In the current experiment, this dimension is represented by a sequential, a branching and a hierarchical arrangement. In the sequential arrangement, typical of program listings and PDL, the control flow and the nesting levels of a program are represented vertically. In the branching arrangement, typical of flowcharts, the control flow is represented vertically and the nesting levels are represented horizontally. Finally, in the hierarchical arrangement, the control flow is horizontal while nesting levels are represented vertically.

This chapter describes three experiments designed to evaluate programmer performance with the various documentation formats. Experiments 1, 2, and 3 involved coding, debugging, and modification tasks, respectively.

Three programs of varying types were chosen for use in these experiments. A program which calculated the trajectory of a rocket was chosen as representative of an engineering algorithm. An inventory system for a grocery distribution center represented the class of programs that manipulate data bases. A third program combined these two types of applications. This program interrogated a data base for information concerning the traffic pattern at an airport and simulated future needs using a queuing algorithm.

These three programs were based on algorithms contained in Barrodale, Roberts, and Ehle (1971). The algorithms were modified to incorporate only the constructs of sequence, structured iteration, and structured selection. They were then coded in FORTRAN and verified for correctness. Each of the resulting programs contained approximately 50 lines of executable code. In addition a short algorithm (11 lines) was used as a practice program.

The procedure for all three experiments was similar. Prior to the experiment, the participants were given a training session in which they were shown each spatial arrangement and each type of symbology. The experimenter described the control flow for each arrangement using a short program as an example; this program was not seen in the actual experiment. The procedure for using a text editor to edit the programs was also explained in detail. Following the training session, the participants were given a short preliminary exercise to familiarize them with the experimental task. Following the exercise, they were given documentation formats and data dictionaries for each of three programs. Using the text editor, they were asked to perform a programming task at a CRT terminal. An on-line data-collection system recorded all interactions with the editor and the time required for each interaction. An automatic checking procedure informed the participants if the program had been compiled and run successfully, or requested that they continue working until a successful execution had been achieved.

The same experimental design was used across the three experiments. Three of the four types of symbology were factorially combined with the three spatial arrangements to produce nine documentation formats. These nine formats were constructed for each of the three programs, resulting in a total of 27 conditions for each experiment.

Participants received a documentation format for each program. Across the three programs, they saw each type of symbology and each spatial arrangement. A participant, for example, might see the first program presented in sequential normal English, the second program in hierarchical PDL, and the third program in branching ideograms. Appendix A presents examples of the types of documentation for a short program.

Different participants were used for each experiment. The participants were assigned to conditions according to the procedures outlined in Winer (1971) and Kirk (1968). Each of the 27 conditions was used once within a set of nine participants. For this 3^3 randomized block design, a minimum of 36 participants is required to assess all interactions and main effects. Across the 36 participants, each program, symbology, and arrangement was presented first, second and third an equal number of times. At the completion of each experiment, the participants were given a questionnaire in which they were asked to state their preferences for the documentation formats they had experienced. They were also asked about their previous programming experience. Previous experiments by the authors had shown a significant relationship between performance and diversity of experience but not between performance and years of experience (Sheppard, Curtis, Milliman & Love, 1979; Sheppard, Kruesi, & Curtis, 1980).

EXPERIMENT 1: CODING

Experiment 1 was designed to measure the effect of the documentation formats in a coding task. The three types of symbology used were normal English, PDL, and ideograms. These three symbologies were combined with the three spatial arrangements, sequential, branching, and hierarchical.

Method

Thirty-six professional programmers from four different locations participated in this experiment. Thirty-two were General Electric employees; the others worked for the Department of Defense. The participants averaged 5.3 years of professional programming experience and had used an average of four programming languages.

The programmers constructed major sections of code at the middle of each of the three programs. For each program, they were given one of the nine documentation formats, a data dictionary, and a listing of the partially completed code. An identical listing appeared on the CRT screen. About 15 lines were missing from the middle of the code. This section included the most complex decision structures present in the program. Working from the documentation, the participants added and debugged the missing section. The dependent variables were the time to code and debug the program and the number of errors.

Results

Coding Time

The participants required an average of 25 minutes to code and debug a program. This represents the amount of time spent using the text editor (i.e., the total time spent at the terminal less the time for compiling, linking, and running). There were large differences in the average times required to code each of the three programs (Table 1). The inventory program required the least time to complete (18.7 minutes); the airport program required the longest time (29.7 minutes).

The difference among the programs was verified by an analysis of variance ($p < .001$). For this analysis, a logarithmic transformation was performed on the coding times to attenuate the influence of extreme scores and to produce a more normal distribution (Kirk, 1968).

Table 2 presents the average coding times for each combination of symbology and spatial arrangement. The normal English versions required 29.7 minutes to complete, the ideograms required 23.9 minutes

TABLE 1
A Comparison of the Dependent Variables for the Three Programs in a Coding Task
(Experiment 1)

	Program			
Dependent Variable	*Inventory*	*Rocket*	*Airport*	*All Programs*
Mean Time to Complete Coding Task (minutes)	18.7	25.7	29.7	24.7
Mean Number of Semantic Errors	0.5	1.0	3.1	1.5
Mean Number of Control Flow Errors	0.3	0.6	2.5	1.1
Mean Number of Assignment and Variable Errors	0.2	0.4	0.6	0.4

and the PDL required 20.5 minutes. This effect of the type of symbology was highly significant ($p < .001$).

Differences due to the spatial arrangement were considerably smaller. Overall, the effect of spatial arrangement was not significant. There were no significant two-way or three-way interactions. It is interesting to note that the PDL presented in the sequential and branching arrangements led to the fastest times. A pairwise comparison revealed that differences among individual cell means greater than 6.9 minutes are significant for $p < .05$, and differences greater than 9.2 minutes are significant for $p < .01$.

Errors

The nature of the errors made by the participants provides valuable information about the difficulties they encountered in coding each program. No error analysis was attempted on data obtained prior to the first submission of a program. For programs that did not compile and run successfully on the first submission, the participants' editing activities for subsequent submisions were analyzed in detail to determine the nature of the errors.

The errors were assigned to two general categories: syntactic and semantic. The syntactic category included a variety of errors that were relatively easy to detect and correct. Unlike the semantic errors, the syntactic errors could be corrected without reference to the specifications. The syntactic errors were few in number and were not related to any particular experimental conditions. Thus, they are of less interest than the semantic errors. The latter were divided into two subcategories:

TABLE 2
Mean Time to Code and Debug in Minutes (Experiment 1)

| | Type of Symbology | | | |
Spatial Arrangement	Normal English	Ideograms	Program Design Language	Total
Sequential	31.8	26.6	16.5	25.0
Branching	26.0	24.7	16.8	22.5
Hierarchical	31.4	20.3	28.1	26.6
Total	29.7	23.9	20.5	24.7

(a) control flow errors, and (b) assignment and variable errors. Control flow errors included the use of incorrect logical operators and transfer of control to the wrong area of a program. Assignment and variable errors included the use of incorrect constants, variables and arithmetic operators.

Table 1 shows that there were large differences in the number of semantic errors across the three programs. These differences follow the pattern previously shown for the coding times. The inventory program was constructed with the fewest errors and the airport program showed the greatest number of errors. It is interesting to note that these differences reside almost entirely in the number of control-flow errors. The airport program resulted in considerably more of these errors (2.5) than either the inventory (.3) or the rocket (.6) program. Thus, the greater difficulty shown by the participants in coding the airport program resulted from problems with the control flow and not with assignment statements or variables.

A similar result occurred when the errors were examined as a func-

TABLE 3
Mean Number of Control Flow Errors (Experiment 1)

| | Type of Symbology | | | |
Spatial Arrangement	Normal English	Ideograms	Program Design Language	Total
Sequential	2.3	2.0	0.2	1.5
Branching	0.3	0.3	0.4	0.3
Hierarchical	2.5	0.9	1.2	1.5
Total	1.7	1.1	0.6	1.1

TABLE 4
Mean Number of Assignment and Variable Errors (Experiment 1)

| Spatial Arrangement | Type of Symbology | | | |
	Normal English	Ideograms	Program Design Language	Total
Sequential	.9	.2	.1	.4
Branching	.7	.6	.2	.5
Hierarchical	.4	.2	.3	.3
Total	.7	.2	.3	.4

tion of the symbology and spatial arrangement. Differences among the nine formats were more pronounced for the control-flow errors than for the assignment and variable errors. As shown in Table 3, the branching arrangement resulted in fewer control-flow errors than the sequential or hierarchical arrangements. It is also interesting to note, however, that a very small number of errors occurred with the sequential PDL format. The assignment and variable errors were more evenly distributed across the nine formats (Table 4).

Preferences for Type of Symbology and Spatial Arrangement

Across the three programs, each participant received documentation in each type of symbology and in each spatial arrangement. On the questionnaire, they were asked to state which symbology and which arrangement they preferred. Table 5 shows these preferences.

In terms of the type of symbology, the majority of participants chose

TABLE 5
Percent of Preferences for Symbology and Spatial Arrangement

| Factor | Experiment | | |
	1	2	3
Type of Symbology:			
PDL	59	33	50
Ideograms	35	34	
Normal English	6	33	18
Abbreviated English			32
Spatial Arrangement:			
Branching	62	58	50
Sequential	23	24	24
Hierarchical	15	18	26

the PDL, ideograms were intermediate, and normal English was the least preferred. In terms of the spatial arrangement, branching was the most preferred, sequential was intermediate, and hierarchical was the least preferred.

Experiential Factors

At the conclusion of the experimental session, the participants were given a questionnaire in which they were asked to report the number of years they had programmed professionally and the number of programming languages they knew. In contrast to our earlier research results, no correlation was found between coding time and these experiential factors.

EXPERIMENT 2: DEBUGGING

In Experiment 2, a debugging task was used to investigate the effects of the documentation formats on programmer performance. The three types of symbology and the three spatial arrangements employed in Experiment 1 were also employed in Experiment 2.

Method

Thirty-six professional programmers from two different General Electric locations participated in this experiment. The participants averaged 6.2 years of professional programming experience and had used an average of 5 programming languages.

The experimental task required finding all the errors in a program and amending the program until it ran correctly. For each experimental program, the participants received a correct version of the documentation. In addition, they received a data dictionary and identical listings of the error-seeded code on the CRT screen and on a paper printout. The practice program was modified to contain one error. The experimental programs each contained three errors. The errors were selected from among errors made in the coding experiment, which had used the same experimental materials. The errors included both transfer of control and assignment/variable errors but did not include syntax errors.

The participants were told that there were several errors in each experimental program and that all of them were located in a predefined, central section of the code. They were instructed to compare the documentation to the code, locate the errors and correct them. If a participant tried running the program without making any changes, the program compiled successfully but produced the message that the output was incorrect.

Results

Debugging Time

The participants required an average of 16 minutes to debug a program. There were no differences among the debugging times for the three programs. The inventory, rocket and airport-scheduling programs required an average of 16, 15.7, and 15.8 minutes, respectively. There was, however, a significant difference among the types of symbology ($p < .05$). The normal English versions required an average of 18.7 minutes compared to 14.5 minutes for the PDL and 14.2 minutes for the ideograms (Table 6).

The effect of the spatial arrangement was not significant, and there were no significant interactions.

Preferences for Type of Symbology and Spatial Arrangement

When asked which symbology and spatial arrangement they preferred, the participants preferred the three types of symbology equally often. In terms of the spatial arrangement, branching was again the most preferred, sequential was intermediate, and hierarchical was the least preferred (Table 5).

Experiential Factors

The number of years the participants had programmed professionally and the number of programming languages they had used were compared to their performance. No correlation was found between years of experience and time to debug. The number of languages used

TABLE 6
Mean Time to Debug in Minutes (Experiment 2)

Spatial Arrangement	Type of Symbology			
	Normal English	Ideograms	Program Design Language	Total
Sequential	19.8	18.2	12.1	16.7
Branching	18.2	14.6	14.6	15.8
Hierarchical	18.1	9.8	16.7	14.9
Total	18.7	14.2	14.5	15.8

and time to debug were correlated, $-.26$ ($p < .06$), indicating that programmers who had experience with a greater number of programming languages required less time to successfully complete the tasks in this experiment.

EXPERIMENT 3: MODIFICATION

In Experiment 3 a modification task was used to compare the performance of programmers using different documentation formats. This experiment employed two of the types of symbology used in the previous experiment, normal English, and PDL. In Experiments 1 and 2, the PDL and ideogram versions were associated with substantially shorter response times than the normal English. It appeared likely that at least part of this difference was due to the manner in which the variable names were expressed. The PDL and ideograms contained the variable names as they were used in the FORTRAN code, while the normal English contained a multiple-word description of each variable. Thus, the PDL and ideograms required less translation from the documentation to the code. An abbreviated English version replaced the ideograms as the third type of symbology in this experiment. This version was included to assess the extent to which the variable names accounted for the shorter response times with the PDL. The abbreviated English was identical to the normal English with the exception that the variable names were used rather than the multiple-word descriptions of each variable. The abbreviated English was more succinct than the normal English but less succinct than the PDL. The same three spatial arrangements (sequential, branching, hierarchical) were again employed in this experiment.

Method

Thirty-six General Electric programmers from three locations participated in this experiment. The participants averaged 8.5 years of programming experience and had used an average of five programming languages.

The experimental task required adding a minimum of three to five additional lines of code to each of three programs. For each program, the participants received a one-paragraph description of the modification, a data dictionary, and a documentation format for the original (unmodified) program. The original code appeared on the CRT screen. The participants were told to make handwritten modifications on the documentation sheets before entering their modification at the terminal. This instruction was given in order to assure that the documentation formats were actually used by the participants.

TABLE 7
A Comparison of the Dependent Variables for the Three Programs in a Modification Task (Experiment 3)

Dependent Variable	Program			
	Inventory	Rocket	Airport	All Programs
Mean Time to Complete Modification (minutes)	21.2	24.9	33.6	26.6
Mean Number of Semantic Errors	0.8	1.2	1.2	1.1

Results

Modification Time

The participants required an average of 27 minutes to modify and debug a program. This represents the amount of time spent studying the program, modifying the documentation format, and using the text editor.

As in the coding experiment, there were large differences in the average times required to complete the modifications for the three programs ($p < .001$). Table 7 shows that the inventory program required the least time to complete (21.2 minutes) and the airport program required the longest time (33.6 minutes).

Table 8 presents the modification times for each combination of symbology and spatial arrangement. In contrast to the previous two experiments differences due to the type of symbology were small. The PDL versions were associated with the shortest performance times, but these differences were not statistically significant.

TABLE 8
Mean Time to Complete Modification in Minutes (Experiment 3)

Spatial Arrangement	Type of Symbology			
	Normal English	Abbreviated English	Program Design Language	Total
Sequential	28.0	28.4	25.5	27.3
Branching	25.4	22.9	21.1	23.1
Hierarchical	30.9	28.6	28.3	29.3
Total	28.1	26.6	25.0	26.6

A significant effect for spatial arrangement did occur ($p < .05$). The branching versions required an average of 23.1 minutes, while the sequential and hierarchical versions averaged 27.3 and 29.3 minutes, respectively. There were no significant interactions.

Errors

The errors made by the participants provide insight into the difficulties encountered when making the modifications. Programs that did not compile and run successfully the first time were analyzed to determine what errors were present in the initial attempt to make the modification. Again, syntactic errors were relatively few in number and were not related to any particular experimental conditions.

The number of semantic errors for each program is shown in Table 7. The inventory program had fewer errors than the other two programs. A detailed analysis of the errors for the inventory program revealed that most of these errors (66%) resulted from problems in placing the statements in the correct locations within the program. The airport and rocket programs were associated with a wider variety of errors.

The pattern of errors associated with the symbology and spatial arrangement was similar to the pattern for the modification times. The effects of symbology were not pronounced, and the branching spatial arrangement was superior to the sequential and hierarchical arrangements (Table 9).

Preferences for Type of Symbology and Spatial Arrangement

Half of the participants preferred the PDL. Abbreviated English was the intermediate choice and normal English was the least preferred (Table 5). As in the previous experiments, the branching spatial arrangement was preferred twice as often as the sequential or hierarchical arrangements.

TABLE 9
Mean Number of Semantic Errors in a Modification Task (Experiment 3)

| Spatial Arrangement | Type of Symbology | | | |
	Normal English	Abbreviated English	Program Design Language	Total
Sequential	1.3	1.8	0.8	1.3
Branching	0.5	0.4	1.1	0.7
Hierarchical	0.8	1.6	1.2	1.2
Total	0.9	1.3	1.0	1.1

Experiential Factors as Predictors of Performance

Two factors relating to programming experience were again compared to the participants' performance times on the experimental tasks. As before, the number of years of programming experience was not correlated with performance, but there was a significant correlation between performance and the number of languages used by the participants ($-.37$, $p < .02$). Thus, programmers who had used more languages completed the tasks in this experiment more quickly.

DISCUSSION

This series of experiments was designed to evaluate the separate and combined effects of type of symbology and spatial arrangement of documentation formats. We found a clear effect for type of symbology. In the coding and debugging experiments, strong differences appeared between the normal English and the PDL. The more succinct symbology, the PDL, was associated with better performance than the more verbose symbology, the normal English. Ideograms were in-between. Although not statistically significant, differences among the types of symbology in the modification experiment reflected the same trends. Further, the novel symbology introduced in the modification experiment, the abbreviated English, was less verbose than the normal English but more verbose than the PDL. Performance times for the abbreviated English fell between the times for the normal English and the PDL, thus reinforcing the conclusion that the more succinct the symbology, the more quickly the programming task will be completed.

Further evidence to confirm this conclusion is supplied by an experiment preliminary to the experiments reported in this paper (Sheppard, Kruesi, & Curtis, 1980). In a comprehension task employing the same documentation formats that were used in the coding and debugging experiments, 72 programmers responded to questions that required tracing through the control flow of the programs. The normal English was significantly slower than the PDL, thus indicating the validity of this result in yet another type of experimental task.

Had the normal English been written casually, one could hypothesize that it was incomplete and misleading. However, the normal English in these experiments was developed very precisely. Assignment, selection, and iteration statements were translated from the original code into the four types of symbology according to a rigid set of rules to insure that the normal English documentation was as complete and precise as the abbreviated English, PDL, and ideograms. It is reasonable to conclude, therefore, that the differences were due to real differences among the types of symbology rather than to an experimental artifact. In real systems development, this effect may be even more pronounced since it is not likely that the documentation is as carefully developed. The combined results of these experiments present strong evidence that detailed program documentation should be presented in a more succinct symbology than normal English.

Evidence regarding the selection of a spatial arrangement for documentation formats is not as strong as for type of symbology. However, the performance times and errors in the modification experiment indicated that the branching spatial arrangement was considerably better for modification tasks than the other two arrangements. A similar, although weaker, result appeared in the coding experiment. The branching spatial arrangement appeared to be mildly superior to the sequential and hierarchical arrangements, particularly in reducing the number of errors related to the control flow.

In the preliminary research cited above (Sheppard et al., 1980), the branching arrangement had been shown to be superior for questions that required hand tracing through the program logic. This is not a surprising result. Programmers have normally used flowcharts, the branching ideogram format, to solve complicated control-flow problems. The interesting result is that the branching arrangement can be combined with another symbology, PDL, to produce a format that can compete favorably with flowcharts.

The participants' preferences for a spatial arrangement agreed with their performance. At least half of the programmers in each experiment preferred the branching spatial arrangement. The hierarchical arrangement was preferred least often, possibly because it was not as familiar to the participants.

The three experiments reported here each produced slightly different results, depending on the experimental tasks. However, performance with the individual combinations of symbology and spatial arrangement was surprisingly consistent from experiment to experiment. Having concluded that PDL is the superior symbology and that the branching arrangement is the preferred arrangement, it was not surprising to see that the branching PDL format was associated with excellent performance in all of the experiments. However, the sequential PDL was also a very usable documentation format. Sequential PDL, the normal arrangement for PDL, has the advantage of being easy to process automatically without the aid of a graphics tool. The sequential, normal English was associated with poor performance in all the experiments.

Strong differences were observed among the three programs used in the coding and modification experiments. The inventory control program was associated with the shortest times and fewest errors, the airport scheduling program resulted in the longest times and most errors, and the rocket trajectory program was in-between. No difference in performance across the three programs was observed in the debugging experiment. One possible explanation for these differing results is that debugging a program from detailed documentation which is known to be correct does not require as much knowledge of the intricacies of the algorithm as does coding or modifying from the documentation. The inherent difficulty of the algorithm may be less important in this type of debugging task than in coding and modification tasks.

It is always interesting to try to relate programmer performance to past experience. In these experiments we compared performance to two experiential factors, number of years of programming experience and number of programming languages known. In all three experiments we found that number of years of programming experience was unrelated to performance. However, number of languages known was correlated with performance in the debugging ($r = -.26, p < .06$) and modification experiments ($r = -.37, p < .02$). The more languages known, the more quickly the programming tasks were performed. Thus, diversity of experience, in terms of the number of languages used, was a better predictor of performance than years of experience. This result replicates results from our previous research (Sheppard et al., 1980; Sheppard, Curtis, Milliman, & Love, 1979) and highlights the importance of ensuring that programmers have an opportunity to gain broad programming experience as part of their professional development.

ACKNOWLEDGMENTS

The authors would like to acknowledge the contribution of Bill Curtis in initiating this program of research. We would also like to thank the following people and organizations for providing participants and facilities: Jack McKissick of GE's Military Electronics Systems Operation, Sue Hannon of GE's Lanham Center Operation, Carol Kiefer, Leo Pompliano, Gene Poggenburg, Len Johnson, Charlie Burns, Jim Coughlin, and Jim Sanchack of GE's Military and Data Systems Operation, Joan Carter of GE's Computer Management Operation, O. J. Barber and Roy Baessler of GE's Industrial Control Facility, and Joan Shields of the U.S. Air Force. The authors would also like to thank Dave Morris and Pete McEvoy for designing the automatic data collection system, Paul Chase for statistical advice, John O'Hare for advice throughout this series of experiments, and Tom McDonald for preparing materials and statistical analyses. This research program was supported by the Office of Naval Research, Engineering Psychology Programs (Contract #N00014-79-C-0595). The views expressed in this chapter, however, are not necessarily those of the Office of Naval Research or the Department of Defense.

APPENDIX A

A SAMPLE PROGRAM USED FOR ILLUSTRATION PURPOSES

This program was not used in the experiments.

**SEQUENTIAL
NORMAL
ENGLISH**

Program to Simulate Waiting Time at a Gas Station

Set the initial values for the random number generator to zero. Read the number of minutes for the simulation from the file 'INFILE'. Do steps 1 through 5 while the number of minutes for the simulation is larger than zero.

1. Set the following variables to zero: the maximum waiting time, the accumulated waiting time, and the number of customers. Set the simulation time to one minute.
2. Do steps A through C while the simulation time is less than or equal to the number of minutes for the simulation.
   (A) If the maximum waiting time is greater than zero, decrease it by one.
   (B) If the number returned by the random number generator is greater than or equal to 0.9, increase the number of customers by one, increase the accumulated waiting time by the maximum waiting time, and increase the maximum waiting time by eight minutes.
   (C) Increase the simulation time by one minute.
3. Calculate the average waiting time by dividing the accumulated waiting time by the number of customers.
4. Print the average waiting time.
5. Read the number of minutes for the next simulation from the file 'INFILE'.

This completes the process necessary to simulate waiting time at a gas station.

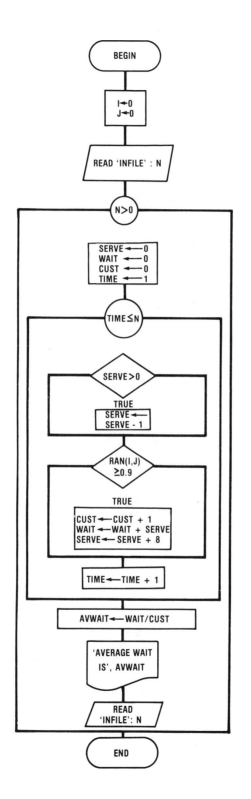

BEGIN

I ← 0
J ← 0

READ 'INFILE' : N

N > 0

SERVE ← 0
WAIT ← 0
CUST ← 0
TIME ← 1

TIME ≤ N

SERVE > 0

TRUE
SERVE ← SERVE - 1

RAN(I,J) ≥ 0.9

TRUE
CUST ← CUST + 1
WAIT ← WAIT + SERVE
SERVE ← SERVE + 8

TIME ← TIME + 1

AVWAIT ← WAIT/CUST

'AVERAGE WAIT IS', AVWAIT

READ 'INFILE': N

END

153

```
PROGRAM GAS STATION

SET I = 0
SET J = 0

READ FROM 'INFILE' : N

DO WHILE N>0

   SET SERVE = 0
   SET WAIT = 0
   SET CUST = 0
   SET TIME = 1

   DO WHILE TIME≤N

     IF SERVE>0

       THEN

         SET SERVE = SERVE - 1

       ENDIF

       IF RAN(I,J)≥0.9

         THEN

           SET CUST = CUST + 1
           SET WAIT = WAIT + SERVE
           SET SERVE = SERVE + 8

         ENDIF

       SET TIME = TIME + 1

     ENDDO

     SET AVWAIT = WAIT/CUST

     PRINT 'AVERAGE WAIT IS', AVWAIT

     READ FROM 'INFILE': N

   ENDDO

   END OF GAS STATION
```

Program to Simulate Waiting Time at a Gas Station

Set I and J to zero.
Read *N* from the file 'INFILE'.
Do steps 1 through 5 while *N* is larger than zero.

1. Set the following variables to zero: SERVE, WAIT and CUST. Set TIME equal to one.
2. Do steps A through C while TIME is less than or equal to N.
   (A) If SERVE is greater than zero, decrease it by one.
   (B) If RAN(I,J) is greater than or equal to 0.9, increase CUST by one, increase WAIT by SERVE and increase SERVE by 8.
   (C) Increase TIME by one.
3. Calculate AVWAIT by dividing WAIT by CUST.
4. Print AVWAIT.
5. Read N from the file 'INFILE'.

This completes the process necessary to simulate waiting time at a gas station.

HIERARCHICAL
NORMAL ENGLISH

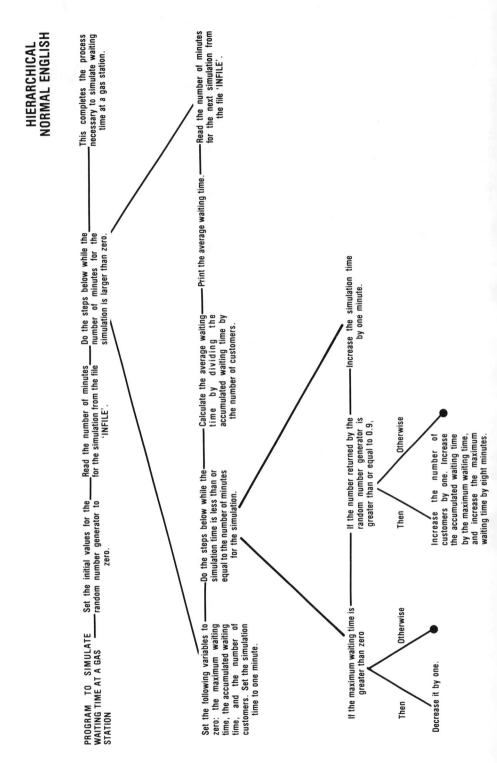

PROGRAM TO SIMULATE WAITING TIME AT A GAS STATION ——— Set the initial values for the random number generator to zero. ——— Read the number of minutes for the simulation from the file 'INFILE'. ——— Do the steps below while the number of minutes for the simulation is larger than zero. ——— This completes the process necessary to simulate waiting time at a gas station.

Set the following variables to zero: the maximum waiting time, the accumulated waiting time, and the number of customers. Set the simulation time to one minute. ——— Do the steps below while the simulation time is less than or equal to the number of minutes for the simulation. ——— Calculate the average waiting time by dividing the accumulated waiting time by the number of customers. ——— Print the average waiting time. ——— Read the number of minutes for the next simulation from the file 'INFILE'.

If the maximum waiting time is greater than zero ——— If the number returned by the random number generator is greater than or equal to 0.9, ——— Increase the simulation time by one minute.

Then

Otherwise

Decrease it by one.

Then

Otherwise

Increase the number of customers by one. Increase the accumulated waiting time by the maximum waiting time, and increase the maximum waiting time by eight minutes.

HIERARCHICAL IDEOGRAMS

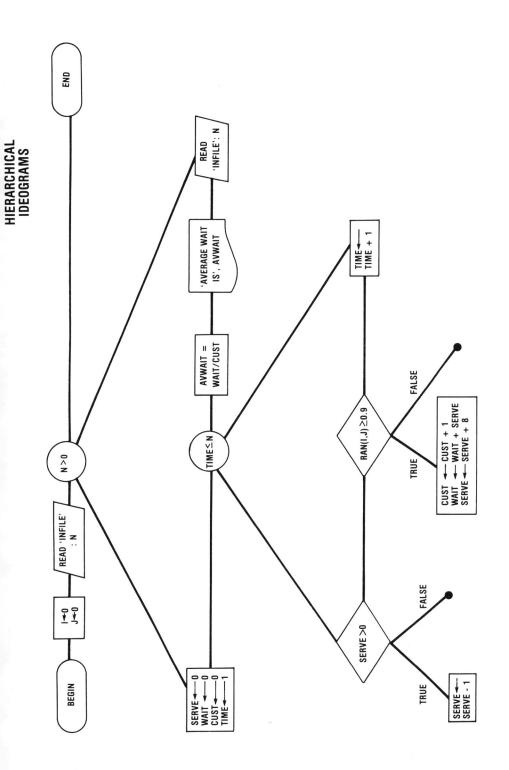

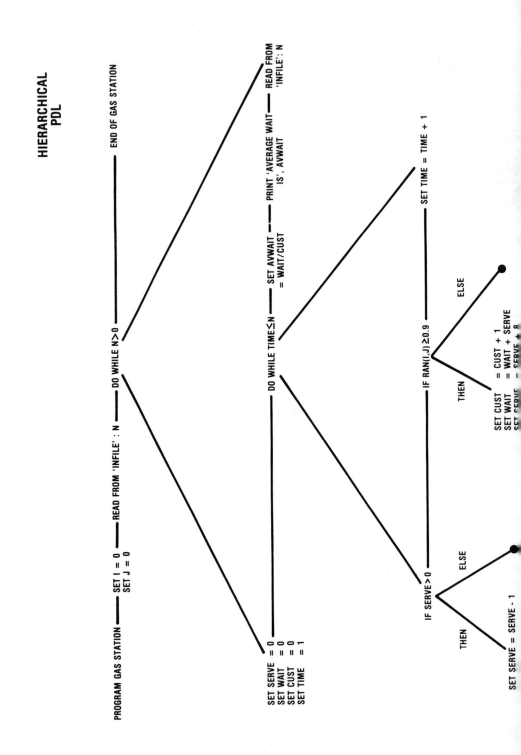

HIERARCHICAL PDL

HIERARCHICAL
ABBREVIATED
ENGLISH

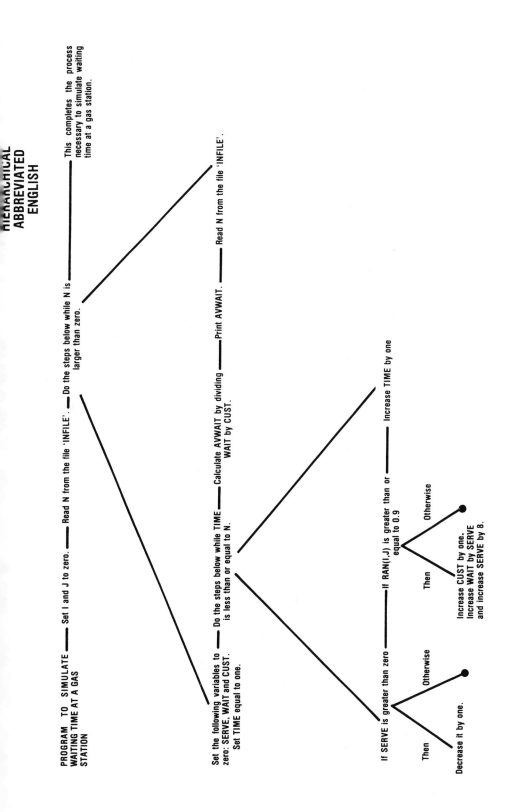

PROGRAM TO SIMULATE
WAITING TIME AT A GAS
STATION

Set the initial values for the
random number generator to
zero.

Read the number of minutes
for the simulation from the file
'INFILE'.

Do the steps to the right while
the number of minutes for the
simulation is larger than zero.

Set the following variables to
zero: the maximum waiting
time, the accumulated waiting
time, and the number of
customers. Set the simulation
time to one minute.

Do the steps to the right while
the simulation time is less than
or equal to the number of
minutes for the simulation.

If the maximum waiting time is
greater than zero

Then	Otherwise

Decrease it by one.

If the number returned by the
random number generator is
greater than or equal to 0.9,

Then	Otherwise

Increase the number of
customers by one. Increase
the accumulated waiting time
by the maximum waiting time,
and increase the maximum
waiting time by eight minutes.

Increase the simulation time
by one minute.

Calculate the average waiting
time by dividing the
accumulated waiting time by
the number of customers.

Print the average waiting time.

Read the number of minutes
for the next simulation from
the file 'INFILE'.

This completes the process
necessary to simulate waiting
time at a gas station.

160

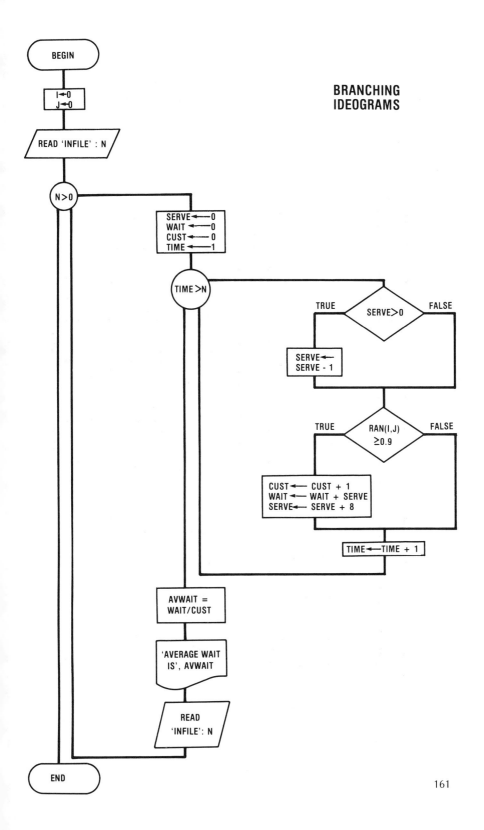

BRANCHING IDEOGRAMS

161

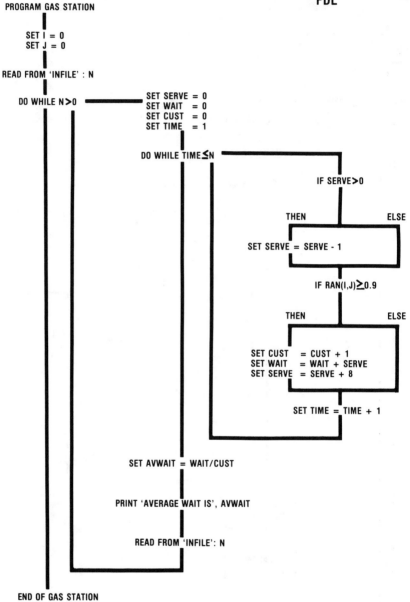

PROGRAM TO SIMULATE
WAITING TIME AT A GAS
STATION

**BRANCHING
ABBREVIATED
ENGLISH**

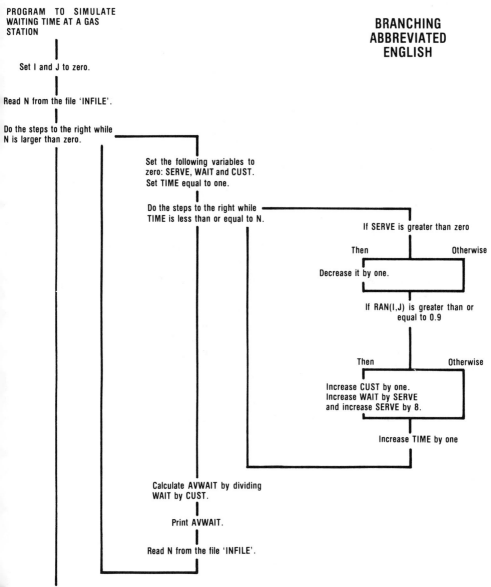

Set I and J to zero.

Read N from the file 'INFILE'.

Do the steps to the right while
N is larger than zero.

Set the following variables to
zero: SERVE, WAIT and CUST.
Set TIME equal to one.

Do the steps to the right while
TIME is less than or equal to N.

If SERVE is greater than zero

Then   Otherwise

Decrease it by one.

If RAN(I,J) is greater than or
equal to 0.9

Then   Otherwise

Increase CUST by one.
Increase WAIT by SERVE
and increase SERVE by 8.

Increase TIME by one

Calculate AVWAIT by dividing
WAIT by CUST.

Print AVWAIT.

Read N from the file 'INFILE'.

This completes the process
necessary to simulate waiting
time at a gas station.

REFERENCES

Barrodale, I., Roberts, F. D. K., & Ehle, B. L. *Elementary computer applications in science, engineering, and business.* New York: Wiley, 1971.

Blaiwes, A. S. Formats for presenting procedural instructions. *Journal of Applied Psychology,* 1974, *59,* 683–686.

Bohl, M. *Flowcharting techniques.* Palo Alto, CA: Science Research Associates, 1971.

Brooke, J. B. & Duncan, K. D. Experimental studies of flowchart use at different stages of program debugging. *Ergonomics.* 1980, *23,* 1057–1091.

Kammann, R. The comprehensibility of printed instructions and the flowchart alternative. *Human Factors,* 1975, *17,* 183–191.

Katzen, H. *Systems design and documentation: An introduction to the HIPO method.* New York: Van Nostrand Reinhold, 1976.

Kirk, R. E. *Experimental design procedures for the behavioral sciences.* Belmont, CA: Brooks-Cole, 1968.

Miller, L. A. Natural language programming: Styles, strategies, and contrasts. *IBM Systems Journal,* 1981, *20,* 184–215.

Ramsey, H. R., Atwood, M. E., & Van Doren, J. R. *A comparative study of flowcharts and program design language for the detailed procedural specification of computer programs.* (Tech. Rep. #SAI-78-078-DEN). Denver, CO: Science Applications, 1978.

Sheppard, S. B., Curtis, B., Milliman, P., & Love, T. Modern coding practices and programmer performance. *Computer,* 1979, *12,* (12), 41–49.

Sheppard, S. B., Kruesi, E., & Curtis, B. *The effects of symbology and spatial arrangement on the comprehension of software specifications.* (Tech. Rept. TR-80-388200-2). Arlington, VA: General Electric, Information Systems Programs, 1980.

Shneiderman, B., Mayer, B. R., McKay, D., & Heller, P. Experimental investigations on the utility of detailed flowcharts in programming. *Communications of the ACM,* 1977, *20,* 373–381.

Winer, B. J. *Statistical principles in experimental design.* New York: McGraw-Hill, 1971.

Wright, P. & Reid. F. Written information: Some alternatives to prose for expressing the outcomes of complex contingencies. *Journal of Applied Psychology,* 1973, *57,* 160–166.

7

A Multilevel Menu-Driven User Interface:
Design and Evaluation Through Simulation

RICKY E. SAVAGE
JAMES K. HABINEK

The design and evaluation of a menu-driven user interface for a general purpose system is described. Analysis of errors made by participants in simulation studies of the interface led to the development of hypotheses concerning user choice behavior. For example, novice users had difficulty selecting menu options based on job titles rather than tasks and functions. Redesign of the interface to reflect these hypotheses resulted in significantly improved performance. The current version of the interface appears to accommodate both the novice user, through an extensive hierarchy of menus, and the experienced user, through a variety of shortcuts to system functions.

People are accustomed to communicating with one another by means of "natural" languages. Given a choice, many people would prefer to communicate with computers in this fashion as well, since natural languages are well-practiced, flexible, and powerful. With a natural language system, the user would not have to remember a large number of idiosyncratic command and procedure names, nor frequently refer to a manual, nor undergo specialized training. However, such systems are not currently feasible, and some debate exists about whether they are really desirable (Ramsey & Atwood, 1979; Shneiderman, 1980).

Many practical alternatives for user-computer communication exist (Martin, 1973; Ramsey & Atwood, 1979). All have advantages and drawbacks, usually as a function of the characteristics of the typical operator. At one extreme, programming languages with precise syntax and vocabularies may be used for dialog between person and computer. These languages have the advantage that complex concepts can be communicated unambiguously by the operator. The associated drawback, on the other hand, is that the use of the language requires extensive training and strict adherence to rules.

At the other extreme, a hierarchy of detailed menus may be used for person– computer interaction. Since this technique relies heavily on the user's recognition memory and passive response to computer prompts, little formal training is required. As might be expected, menus have their drawbacks, too. For example, the creator of the menu must have a perception of all the possible or desirable options to include. Since this set of options cannot be infinite, limitations in the actions possible through the use of menus are unavoidable. Furthermore, use of the menus may become tedious if the choices are too finely detailed or if the user has extensive experience.

Somewhere between the extremes of programming languages and menu selection lie techniques such as form-filling or fill-in-the-blanks. These have the disadvantage that some training is necessary to teach users the acceptable responses, but they have the advantages of eliminating lengthy displays of options and of permitting more than one response in a single interaction.

The popularity of general purpose systems has also created human–computer interface problems. The very nature of general purpose systems, such as multiple display stations and a wide range of users, presents unique problems. The typical interface to these systems is a cryptic command language requiring multiple parameters. These commands and their associated parameters are difficult to remember (over taxation of recall memory) and difficult for novice or casual users to use. To make adequate use of a typical command language interface, constant reference to a manual is needed which tends to be time consuming and often frustrating. As the number of system users increases, the degree of formal training of the typical user declines. Techniques, such as menu selection, which can best accommodate the novice user, almost necessarily must be included in a strategy for person–computer communication. Yet, care must be taken that the experienced or sophisticated user is not encumbered with a system that involves frustratingly slow entry of commands or procedures.

The purpose of the current research was to design and evaluate a user interface that would satisfy a broad spectrum of users, from competent system programmers to complete novices. The goal was to produce an interface that would easily guide the first-time user through a series of menus to the desired procedure, but at the same time provide the experienced user with the flexibility to take time and effort saving shortcuts. This chapter describes in detail the process and techniques required to develop and test an interface that would satisfy the needs of a broad spectrum of users. Two design and evaluation interations are described.

DESIGN OF THE INTERFACE

Background

In order to induce realism in this effort, the IBM System/34 was chosen as the system with which to design and evaluate this interface. The System/34 is a general purpose data processing system designed for a wide range of applications. It supports up to 16 display stations locally and up to 64 display stations via communication lines at remote locations. This system can be a processor terminal, a host system, or a sub-host system in a data communication environment. The data processing experience of the System/34 users ranges from no data processing knowledge to full-time programming.

Work is accomplished on the System/34 with a series of commands and procedures as well as an Operator Control Language (OCL) with which the operator can create his own procedures. Since commands and procedures are limited to eight characters, it is difficult to assign them meaningful names and, therefore, they can be difficult to remember. Likewise, each command and procedure can have many parameters which can also be difficult to remember. To alleviate some of these problems, System/34 provides a tool which allows the user to build menus so he does not have to know the command or procedure name. In addition, a help procedure is provided which can be called to display prompts that contain the syntax, default parameters, keywords, and one-line descriptions for the specific commands and procedures. The primary use of the System/34 Help function is to assist the user with the syntax of a command rather than to provide him with assistance in choosing the command needed to perform a task.

To illustrate the System/34 Help function, assume you want to create or build a library to store a program, but cannot remember the command name. You press the "HELP" key and the display in Figure 1 appears. An immediate problem occurs if you do not know what type of command you need. In this example, choosing Option 1 is the correct solution and the display screen in Figure 2 appears. This display screen is already busy, but by using the "ROLL" key, even more procedure commands appear. The brief two- or three-word description for each procedure command is often inadequate. In this example, the correct procedure command is "BLDLIBR," and if you key this in, the display screen in Figure 3 appears. This display screen prompts you for the various parameters required to run this procedure command. No help text exists to explain the meaning of these parameters.

This example illustrates the many shortcomings of most general pur-

```
                      SYSTEM/34 'HELP' CATEGORIES


    1.  SSP Procedure Commands

    2.  Control Commands

    3.  Operation Control Language--OCL

    4.  Procedure Control Expressions

    5.  Utilities--DFU, SDA, SEU, SORT, WSU

    6.  Languages and Compilers--ASM, BASIC, COBOL, FORTRAN, OLINK, RPG

    7.  Data Communications Procedures

    8.  Service Aids Procedures



    ENTER NUMBER OF CATEGORY ---> 1_          CMD KEY 1 -- FUNCTIONS

                                              CMD KEY 7 -- CANCEL
```

FIGURE 1. **System/34 help categories as a result of pressing "HELP" key.**

pose systems. The user with little or no data processing skill is normally unable to perform this and similar system functions. Even a skilled programmer would usually need to refer to the manual to successfully use these commands. The purpose of this user interface was to overcome these shortcomings and to provide a structured approach which allows

```
                        SYSTEM/34 PROCEDURES


    ALTERBSC  - Alter BSC Parameters     CONDENSE - Free Library Space

    ALTERSDL  - Alter SDLC Parameters    COPYI1   - Copy Diskette

    BACKUP    - Backup System Library    COPYPRT  - Copy From Spool File

    BLDFILE   - Create Disk File         CREATE   - Generate Message Member

    BLDLIBR   - Create User Library      CRESTART - Restart Checkpoint Task

    BLDMENU   - Create Menu              DATE     - Change Session Date

    BUILD     - Correct Disk Data        DELETE   - Delete File or Library

    CATALOG   - Display VTOC             DISABLE  - Disable Subsystem

    COMPRESS  - Free Disk Space          DISPLAY  - Display Data File


    ENTER PROCEDURE NAME TO BE EXECUTED ---> BLDLIBR      ROLL TO CONTINUE
```

FIGURE 2. **System/34 SSP procedures as a result of entering Option 1 from previous display screen (see Figure 1).**

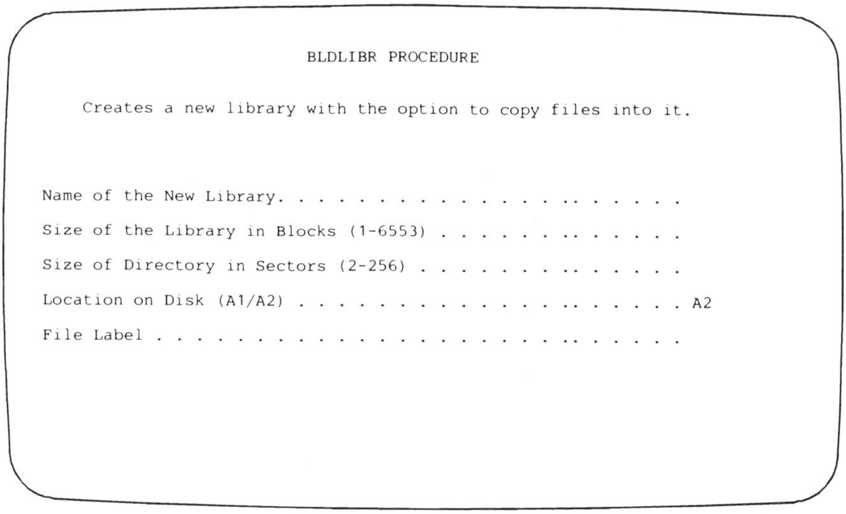

```
                         BLDLIBR PROCEDURE

        Creates a new library with the option to copy files into it.


   Name of the New Library. . . . . . . . . . . . . . . . . . . .

   Size of the Library in Blocks (1-6553) . . . . . . . . . . . .

   Size of Directory in Sectors (2-256) . . . . . . . . . . . . .

   Location on Disk (A1/A2) . . . . . . . . . . . . . . . . . . A2

   File Label . . . . . . . . . . . . . . . . . . . . . . . . . .
```

FIGURE 3. **System/34 prompt display as a result of keying BLDLIBR from previous display screen (see Figure 2).**

users to perform system functions without relying on recall memory, extensive experience or training, or constant reference to a manual.

Initial Human Factors Considerations

The primary consideration for designing this user interface was to provide an easy access to the entire system without constant reference to manuals. The design was targeted to assist the novice user, but at the same time not to penalize the experienced user. This second point is important and to accomplish this goal, systems may have to be designed with multiple levels of user interface. General purpose systems ordinarily have all levels of users—from novice to experienced programmer. For experienced users, there should be a highly abbreviated and quick access to system functions. The novice user, on the other hand, needs a very specific step-by-step interface to lead him to the required system function. Both Shneiderman (1980) and Martin (1973) have stated that novice users normally require a menu driven interface.

A consideration in designing the current interface was the structure of the menus. The hierarchical or tree structure was used (see Shneiderman, 1980, Chap. 7) because of the experimental evidence favoring such a structure (Brosey and Shneiderman, 1978), and because of the natural hierarchical structure of this interface (Durding, Becker, & Gould, 1977). To illustrate this second point, consider a user who needs to copy a library from diskette. His first choice from the first menu would be the option of "work with libraries." From the second level of menus, he would choose the "copy libraries" option, and finally, from the third level of menus he would choose the "copy libraries from diskette" option. This option would give the user the command to copy a library from diskette. In this example, the user proceeds from a general level to a specific point which shows the hierarchical nature of the task.

Another initial consideration was consistent screen design. Based on the work by Engel and Granda (1975) and Peterson (1979), screen standards were developed to ensure consistency. It was also determined that menus should have no more than nine options and that the paths to a function should be relatively short, generally three of four levels of menus.

In order not to penalize the experienced user, several shortcuts to commands exist. For example, any given menu may be displayed simply by entering its name. If a user needs to work with a diskette, he can key in the keyword "DISKETTE" and a menu is displayed with all the diskette functions. This normally saves the user one to three levels of menus. The use of keywords was designed as a part of the interface. Three other more sophisticated levels exist which are provided by the System/34. One level was described earlier by the use of the "HELP" key (see Figures 1, 2, and 3). A second level of interface involves the user directly entering a known command name to obtain a prompt screen (see Figure 3) for that command. Finally, the command name may be entered with its appropriate positional parameters to most directly accomplish a desired function. This multiple level interface spans the range of users from complete novice (menus) to experienced user (directly entering command names).

The Design Team

The next phase of the design was to determine the nature of the menus: how to step the user to the various functions, how to word the menus, the levels of menus required to reach the function, etc. Human Factors and Programming did the initial design of the menu-driven user interface on paper. Subsequently, a design team was created involving Human Factors, Programming, Market Planning, and Publications. Through the design team, many changes were made in terms of wording, new options on the various menus, and function key changes. Other changes were made by adding and deleting menu items, and modifying the method of going from one menu to another and paging backwards through the menus. The interface was restarted and reorganized several times, primarily because it was difficult for the design team to come to agreement as to what the user needed and what was easy to use. The design committee was limited in its ability to play the role of an operator and to attempt to perceive all the mistakes that were likely to occur. An initial design was agreed to and a working simulation or prototype was developed to allow an evaluation of the interface.

Implementation of the Simulation

Very little supporting software was required since most of the support was already available on the IBM System/34. The Source Entry Utility (SEU) and Screen Design Aid (SDA) were used extensively to create the menus and prompts. System/34 had some restrictions which prohibited running some commands from a menu. These restrictions were removed and code was added to process the command keys and chaining of menus. These operating system changes were primarily modifications to existing System/34 assembler routines.

Description of the Interface

A shortcoming of the System/34 is the inability for the customer with little or not data processing skill to perform occasional data processing work without referring to a manual. This interface provides a structured menu approach to performing work on the system. The menus are worded with as little data processing terminology as possible and are chained together so that the operator can progress forward and backward from one level to the next. The operator can also return to the start of the menu structure by means of a single function key.

To illustrate the use of the menu-driven interface, consider the previous task discussed—the need to build a library. The first menu displayed is shown in Figure 4. This menu leads users to every system function available in a hierarchical manner, from general or broad topics to specific tasks or functions. To build a library, Option 3 should be chosen. The display shown in Figure 5 represents the second level of menus. This menu contains all the tasks dealing with libraries. Option 1 is the correct choice, and Figure 6 shows the prompt display for the needed procedure. This menu approach, when compared with the "HELP" facility on the System/34, requires only that the user understand the task. The user does not need to know command names or parameter values and there is no need to refer to a manual. The complete design of the interface will include extensive help text for every menu as a whole, for every menu option, for every prompt display as a whole, and for every prompted parameter.

To illustrate how the keyword interface works, the user simply keys in

```
                              MAIN MENU


     Select one of the following:

     1.   Select a user menu or user library
     2.   Data file functions
     3.   Library functions
     4.   Diskette and printer functions
     5.   Programming operations
     6.   Work station operations
     7.   System control operations
     8.   Learn about the interface
     9.   Sign-off the system




     CMD 3--Page back                  CMD 7--End

     ENTER OPTION NUMBER

     3_                           <--READY
```

FIGURE 4. **The first menu of the interface.**

the keyword "LIBRARY" and receives the library menu shown in Figure
5. The user then chooses Option 1 and the prompt display in Figure 6
appears. In this example, only one level of menus is saved, but typically
two or three levels of menus can be saved using keywords. The user can
get access to the keyword list by keying in "HELP KEYWORDS."

```
                           LIBRARY FUNCTIONS


     Select one of the following:

     1.   Build a library
     2.   Copy a library
     3.   List library information
     4.   Delete a library
     5.   Change the name of a library
     6.   Update a library





     CMD 3--Page back        CMD 5--Main menu        CMD 7--End

     ENTER OPTION NUMBER

     1_                           <--READY
```

FIGURE 5. **The second level of menus as a result of choosing Option 3 from the previous
display screen (see Figure 4).**

```
                         BLDLIBR  PROCEDURE

        Creates a new library with the option to copy files into it.



      Name of the New Library. . . . . . . . . . . . . . . .. . . . . .

      Size of the Library in Blocks (1-6553) . . . . . . . . . . . . .

      Size of Directory in Sectors (2-256) . . . . . . . . . . . . . .

      Location on Disk (A1/A2) . . . . . . . . . . . . . . . .. . . . . A2

      File Label . . . . . . . . . . . . . . . . . . . . . . . .. . . . .
```

FIGURE 6. **Prompt display as a result of choosing Option 1 from previous display screen (see Figure 5).**

Finally, it is important to note that this interface is flexible and somewhat redundant. The interface is flexible in that the user can always back out from anywhere in the interface. Therefore, the user will never be stuck without an exit. The interface is also redundant in the sense that many functions can be arrived at by more than one path. This redundancy is not immediately apparent to the user.

EVALUATION OF THE USER INTERFACE (PHASE I)

Methodology

Some specific problems had to be solved initially to successfully evaluate the interface. It was relatively obvious that in order to evaluate the user interface, participants would have to perform tasks and use the menus to aid them. But how were the tasks to be developed and selected? Also, how were the data to be collected and what type of participants would be used for the evaluation? From the start, a decision was made not to test any help text or keywords during these evaluations. The evaluation therefore stressed the menu interface itself with no shortcuts (keywords) or any online assistance (help text). The importance of this decision was to ensure the menus were designed properly and could withstand the evaluation before testing help text. It was felt that the availability of help text would tend to cover up possible problems with the menus.

Tasks

With help from individuals with field experience, tasks were developed for programmers, system operators, and work station operators. Careful consideration was given to importance, frequency of use, and difficulty of the many possible functions before they were selected for use. The tasks were pilot-tested with 12 participants to ensure understandability. This was an important step because of possible problems of confounding the lack of understanding the instructions and using the interface. Some examples of tasks were: building libraries and files, compiling and modifying source, and listing the contents of a file or directory for programmers; changing printer ID, establishing user and system security, sending messages, duplicating diskettes, and running jobs for system operators; and copying files to and from diskette, replying to messages, bringing up a user application, and deleting the contents of a diskette for work station operators. Each participant had at least 10 tasks to perform.

Participants

Twelve participants performed the programmer tasks: Eight were programmers with less than six months experince, one participant was a programming librarian, and three participants had some system experience, but were not programmers. Nine participants performed the system operator tasks and 10 participants performed the work station operator tasks. These participants' experience ranged from total novice to operators with less than six months experience. All participants were employees serving in the study voluntarily.

Equipment

The simulation of the user interface was developed to run on the System/34 which, in turn, became the vehicle for running the study. The system captured keying responses from the participants (menu selections, function keys, etc.) and time on each menu for the protocol analysis. Video recorders and cameras were used as a backup and to record comments by the participants.

Procedure

Each participant operated the system individually for up to a two-hour session. General instructions were given and the participants were allowed to ask questions before continuing without experimenter assistance. Participants read the task instructions and used the menus to accomplish their tasks. Manuals were available throughout the session. The participants chose options from a series of menus to lead them to a command, procedure, or utility. Parameters were then entered on a prompt screen from which the function executed and accomplished the task. If participants could not accomplish a particular task, they were allowed to continue with the next task. Finally, at the end of the session, an attitude questionnaire was administered using a five-point rating scale, and participants were subsequently encouraged to comment freely concerning their feelings about the interface.

Data Analysis

The analysis of the user interface posed special problems because the evaluation was not an experimental comparison of alternatives. The analysis procedures had to be sensitive in finding problems, such as the wording of the menus, the hierarchical structure of the menus, and the organization of a specific menu path or a specific menu, and to be sensitive in describing user behavior, particularly user errors.

Time was recorded for each menu and each task. The primary benefit of measuring time was to serve as a baseline for future evaluations (as a result of modifications to the interface) and to determine if an unusually large amount of time was spent on a particular menu or task which could point to potential problems with the interface.

The protocol analysis consisted of mapping each subject's responses on paper for a comparison with the optimal path, which was defined as the shortest route to the desired function. The error analysis consisted of categorizing user errors (defined as an incorrect menu option chosen or an incorrect parameter specified) into types of errors. A probability analysis was performed for each menu. Each menu was looked upon as a decision point and the probability of a correct decision was determined. The probability analysis pointed to specific menus which were not communicating adequately. All of these analyses used the participants' first encounter with a menu to determine an error. The participants often corrected their mistakes and successfully completed the task, but for the purpose of analysis the initial encounter with the menu was used.

The primary benefit of measuring time and error frequency was to serve as a baseline for future evaluations. Changes and modifications were made to the interface as a result of the initial evaluation. The success of these changes was validated by reevaluating the interface and improving upon the measurements: less time required to perform the task and fewer errors made.

Results and Discussion

From the protocol analysis, the error analysis and the probability analysis were performed. The error analysis produced four general categories of errors. The first category was called an *inconvenience error* and resulted from three different actions: (a) the user taking the wrong path, but ending up at the correct function; (b) the user searching or exploring various menu options and paths and eventually taking the correct path to the correction function; and (c) the user searching or exploring and eventually taking the wrong path, but ending up at the correct function. These inconvenience errors were not serious errors, but the user took a less than optimal path to the correct function.

The second category was called a *path error*. Three different actions could cause this error: (a) the user taking the wrong path to the wrong function; (b) the user searching or exploring and taking the wrong path to the wrong function; and (c) the user taking the correct path, but on the last menu selecting the incorrect option which led to the wrong function. The path errors were considered serious because the user ended up at the incorrect command or procedure.

A *function error* was the third category and resulted when the user filled in the wrong parameters on the prompt screen after having successfully reached the correct function. This error did not indicate problems with the menus, but did point out problems the user had with the prompts. Finally, the fourth category was labeled *miscellaneous errors*. These errors consisted of failure to follow instructions, omission of the task, etc., and were not necessarily descriptive of user behavior concerning the interface.

Participants could make more than one type of error per task. Furthermore, a participant could make errors but still complete the task successfully by correcting his/her errors. A failure of a task was defined when a participant committed a path error and ran the wrong function or made an uncorrected function error.

Table 1 shows the distribution of errors, in percent, for each category and for each type of task. The inconvenience errors are relatively consistent for all three types of tasks and they seem to be due to unclear

TABLE 1
The Percentage of Errors for Each Category (Phase I)

Task Type	Inconvenience	Path	Function	Miscellaneous
Programming	26.4	12.8	4.0	56.8
System Operator	31.8	35.0	29.1	4.0
Work Station Operator	24.2	57.8	11.8	6.2

wording of the menu options, lack of understanding of some terminology, and some misinterpretation of how the task should be accomplished. This interface was designed for the System/34 which has its own terminology already defined. Many of the problems users have with wording or terminology are a result of the idiosyncrasies of the System/34. Some of these problems can be "worked around," whereas others will have to be defined and explained in help text. One of the most useful and interesting results of the error analysis was pointing to where specific help text is needed beyond what was already planned. The path errors were rather high for system and work station operator tasks. Discussion of the probability analysis will explain some of these errors. The function errors for system and work station operator tasks seemed to be due to a lack of experience with the command and procedure prompts and a lack of help text explaining what the parameters were, what kind of information was needed, and recommended values.

For programmer tasks, the "miscellaneous errors" category also consisted of improperly backing up through the menus. In these programmer tasks, users were allowed to back up from a successfully completed task to the previous menu (rather than to the first menu in the hierarchy) which would lead them to the next task. This backing up through the menus was allowed for programmers in order to represent the more continuous nature of the programmer's work as opposed to the more discrete nature of the operator's work. Operators would be expected to return to the first menu for the next task. However, all users predominantly backed up to the first menu to begin the next task which was a less than optimal strategy. This behavior accounted for 45% of the total errors which left 11.8% for the miscellaneous category. From interviewing the subjects, two explanations appeared. First, closure (Shneiderman, 1980) seemed to be important in that users had a feeling of accomplishment upon completing one task and backing up to the first menu to begin the next task from the start. Second, users stated that by beginning from the start of the menu hierarchy, they were learning the menu structure. It appears that the users were maintaining a cognitive or mental picture of the menu structure by doing this.

The probability analysis was successful in pointing to menus in which users had a low probability of selecting the correct option. In general, these problem menus were of two types. In the first type, one or possibly two incorrect alternatives had a high probability of selection relative to the correct alternative. These errors appeared to result from a discrimination problem. Figure 7 shows an example of a discrimination problem where the correct option was chosen only 33.3% of the time. The task involved copying a library to a file on diskette (Option 4 in Figure 7), whereas the competing response (Option 3 in Figure 7) involved copying

```
                           COPY A LIBRARY
                                                     Probability
        Select one of the following:               of Selection

        1.   Create a library and copy members to it

        2.   Copy a file to a library

        3.   Copy library members to a file on disk or diskette      .667

        4.   Copy the contents of a library to a diskette file       .333

        5.   Copy library members to another library

        6.   Restore members from diskette

        7.   Save system library on diskette

        8.   Restore system library from diskette
```

FIGURE 7. **Example of a discrimination problem (correct option is 4).**

a member of a library to a file either on disk or diskette. This subtle difference resulted in the discrimination problem. These problems were normally corrected by wording changes to make the competing alternatives more discriminable.

In the other type of problem menu, participants appeared to use a shotgun approach: A large variety of incorrect alternatives was selected by the participants in lieu of the correct option. In this case, the users appeared to have little notion as to which alternative was correct. The path errors for the system and work station operator tasks illustrate this. These two groups of participants were constantly confusing each others options from the first menu (Options 6 and 7 in Figure 4). At the next level, the shotgun approach resulted. An example of the shotgun approach is shown in Figure 8. There were five competing responses. In addition, one third of the participants were so confused they exited from the menu. Only one person selected the correct option on the initial encounter. Most of the shotgun problems appear to occur on menus with a large number of options, which tends to be an information overload. Too many ideas or different classes of functions were put on a single menu. In Figure 8, five different classes of functions exist. Furthermore, the two types of operators appeared not to think in terms of job classification, as shown by the first menu in Figure 4 (Options 6 and 7). The participants wanted to think in terms of the type or class of functions or tasks rather than in terms of whether they were a system versus a work station operator. These two paths needed redesign, resulting in organization by function rather than by type of user.

Table 2 presents the average response to the questionnaire using a five-point scale, with one being positive and five being negative. These results show that the wording of the menus was the major problem. This

```
                        PERFORM SYSTEM CONSOLE OPERATIONS

                                                          Probability
      Select one of the following:                        of Selection

      1.  Control spool print files                            .22

      2.  Control current jobs

      3.  Control jobs on job queue

      4.  Work station commands                                .11

      5.  General status information                           .11

      6.  Communication commands

      7.  Display commands                                     .11

      8.  Control messages

      9.  Other system operations                              .11

          (Used Function Key to Exit)                          .33
```

FIGURE 8. **Example of a shotgun approach (correct option is 4).**

problem is primarily a result of the terminology of the System/34. Some of these words can be worked around, whereas others will have to be defined in help text. Generally, the participants felt that the interface was easy to use, helped them to learn about the system, and would aid in their productivity. The question concerning how the interface helped the participants learn about the system came as a result of interviewing participants during the pilot test and during the programmer tasks. These participants volunteered the comment that by using this interface, they learned something about how the system was structured and organized. This response was an unexpected result, and the questionnaire was expanded to include this question. From Table 2, system operator participants rated this question the most positive aspect of the interface, whereas the work station operators rated it second. The hierarchical structure of this interface appears to complement or structure the user's cognitive model of the system. The earlier discussion of programmer

TABLE 2
Average Response to the Attitudinal Questionnaire (Phase I)

Task Type	Overall Rating	Menu Structure	Menu Wording	Productivity of Work	Ease of Use	Learning System
Programmer[a]	2.33	—	—	—	—	—
System Operator	2.39	2.44	2.67	2.11	2.22	1.94
Work Station Operator	2.60	2.60	3.20	2.20	2.50	2.30

[a]The questionnaire was expanded after these participants had completed their tasks.

participants preferring to back up and start each task from the beginning is also indicative of the user's attempting to form a cognitive model of the system by using this interface.

One of the primary results from this evaluation was the high overall success rate: programmer tasks, 92% system operator tasks, 79%; and work station operator tasks, 81%. Another major result was that the participants did not use the manuals. These results suggested that a menu-driven user interface offers a viable method of assisting users in their work. On the other hand, the fact that failures did occur, and that frequently many false starts and backtracking and much searching and exploring of various menu paths were required before acceptable solutions were found, suggested that the interface needed some redesign. The design of the interface did not always match the user's cognitive model of how the interface should work, as shown by the operators looking for classes of functions rather than job classification. Finally, many of the terminology problems needed to be solved by help text because of the way the System/34 is designed.

REDESIGN AND REEVALUATION (PHASE II)

Many of the changes to the interface involved breaking up and rewording complex and cluttered menus to make them simpler in appearance and to reduce the information overload. As discussed earlier, two menu paths had to be redesigned, which affected some of the structure of the menu hierarchy. These paths originally required users to choose options which corresponded to their job: "work station tasks" or "system console tasks." The results from Phase I clearly showed that users were not inclined to use the menus in this fashion. They looked for task-oriented options rather than job classification options. The redesign eliminated the two job classification options and substituted four task-oriented options: system and user definition, problem determination, communication commands, and system management. As in the Phase I evaluation, no help text was used, with one exception. Given the high number of function errors for operators (see Table 1), help text was included to define and explain the parameters required for the commands. The purpose of the Phase II evaluation was to test the success of these changes.

Method

Twenty people participated in the second phase of testing. Six performed the programmer tasks, seven performed the system console operator tasks, and seven performed the work station operator tasks. The level of participant experience was similar to Phase I. The tasks, procedures, and equipment were identical to Phase I.

Results and Discussions

The second phase of testing the interface was completed with significant improvements in user performance. Table 3 shows the time and success rate for the three groups of participants, comparing the results of the first and second phases of testing. As can be seen, significant improvements were obtained for both types of operators in terms of the average time to complete a task and the rate of successful completion of the tasks.

The results of the error analysis are presented in Table 4. Clearly, the path errors were the most serious, and they were significantly reduced during the second phase of testing for both types of operators. The function errors were also significantly reduced, especially for programmer and system operator participants. This reduction of function errors demonstrates the success of the help text provided for the parameters of the various commands. The frequency of the inconvenience errors remained about the same. Failure to significantly reduce the inconvenience errors was due in part to the following: (a) the lack of clarity of System/34 terminology, which will have to be solved with detailed help text; and (b) the availability of redundant paths (more than one path to a function), most of which are less than optimal. This second problem is not serious in that it allows the user flexibility in how he/she "sees" or interprets the solution to the problem. Exact predictions of the users' cognitive model is often impossible, but allowing this flexibility in choosing alternative strategies and, therefore, alternative menu paths to solve the problem is desirable. Future testing of the interface with help text will determine the success in eliminating the more serious inconvenience errors. However, mismatches of users' cognitive model will remain and so will some of the inconvenience errors. Tables 3 and 4 demonstrate the success of the changes that were made to specifically help the operators.

TABLE 3
Average Time to Complete Each Task and the Success Rate of Both Phases of Testing

Task Type	Test Phase	Time (minutes)	Success
Programmer	1	11.53	92%
	2	12.85	100%
System Operator	1	5.41	79%
	2	1.69	95%
Work Station Operator	1	6.30	81%
	2	2.92	95%

TABLE 4
Average Errors per Participant by Error Category for Both Phases of Testing

Task Type	Test Phase	Error Category		
		Inconvenience	Path	Function
Programmer	1	6.00	1.60	2.20
	2	7.30	2.00	.83
System Operator	1	4.80	5.30	4.40
	2	3.71	.71	.29
Work Station Operators	1	4.33	10.33	2.11
	2	5.71	.43	1.14

The probability analysis revealed several problems, but they were minor compared to those in the first evaluation phase. These problems consisted of some menu options being misleading, missing functions, wording problems, and discrimination problems. The problem of the shotgun approach was virtually eliminated, again demonstrating the success of the changes made to the interface. Generally, all of these problems had obvious solutions and were easily correctable.

Finally, the administration of the attitudinal questionnaire, the results of which are shown in Table 5, produced similar results to Phase I. The wording of the menus was again the major problem, especially for work station operators. Also, the question concerning the interface's ability to help the user learn about the system was again rated the most positively by the participants. The ease of use of the interface was rated very positively, especially by the programmer and system operator participants.

SUMMARY OF THE RESULTS AND FUTURE TESTING

From the Phase II evaluation, it is evident that the changes made to the interface were successful. This success can be seen in terms of average

TABLE 5
Average Responses to the Attitudinal Questionnaire (Phase II)

Task Type	Overall Rating	Menu Structure	Menu Wording	Productivity of Work	Ease of Use	Learning System
Programmer	2.0	2.2	2.5	2.3	1.8	1.5
System Operator	2.4	2.6	2.6	2.6	2.1	2.1
Work Station Operator	2.4	2.4	3.6	2.3	2.6	2.3

time to complete the task, percent of successfully completed tasks, and fewer path errors for the operators. The need for detailed help text is apparent from the lack of reduction in inconvenience errors. The reduction of function errors demonstrates how help text can be a definite aid to user success. Many of the problems found in these evaluations were due to ambiguous terminology. Participants did not know the difference between work station, display station, and device, nor did they understand the difference between saving and copying a file, nor removing and deleting a file. This problem needs to be solved not only with simpler terminology, but also with help text for each menu to explain in more detail what each particular menu option means. In general, the results of these two evaluations illustrate the success of the methodology used in designing and in evaluating the interface. User performance showed significant improvement, and user attitude was consistently positive.

Several results from these evaluations contributed significantly to a better understanding of user behavior with menus. First, it was evident from these studies that a user's job title is not important, whereas a user's tasks and functions are important when developing menus. This was demonstrated by the success of the changes made from Phase I. Second, it appears that users in these studies preferred shorter menus with more levels than the opposite case. Many of the changes from Phase I consisted of breaking up a complex menu into two menus. The second study showed the success of these changes. This result is in apparent conflict with a recent study by Dray, Ogden, and Vestewig (1981). Differences between the studies may be attributed to the realism of this interface (in that it dealt with an actual system) and to the fact that menu options were word phases as opposed to one or two words in the Dray et al. (1981) study.

Third, the results of these evaluations suggest that the hierarchical design of the interface either matched the users' cognitive model of the system or made it easy for the users to form a cognitive model of the system. This notion of the users' cognitive model was supported by the interviews, attitudinal questionnaire, and the strong tendency for programmers to use the less than optimal strategy of starting each task from the beginning. The failure to reduce the inconvenience errors, however, points to some mismatching of the interface and the users' cognitive model. The existence of redundant paths to functions is a positive feature of the interface in that it allows for this mismatch even though a less than optimal path often results. Fourth, closure by the users seems to be important, particularly with programmers, indicating the need for users to complete a task successfully before starting the next task. For the programmer tasks, the need for closure resulted in a less than optimal strategy. Finally, reference manuals were available for all participants during both evaluations. These manuals were never consulted with the exception of a glossary. The problem with the System/34 terminology is again apparent and the help text should solve this problem. Generally, however, the interface successfully demonstrated its ability to alleviate the constant need to refer to a manual.

Future evaluations of this menu-driven interface will involve several goals. One goal is to use more novice participants and even nonemployees. New tasks to validate these evaluations will constitute the second goal. A third goal is to design the help text for the interface for every menu as a whole, every menu option, every prompt for commands, and every prompted parameter for commands. Fourth, an evaluation of the keyword aspect of the interface is needed. Finally, several new methodologies can be explored. Thinking-aloud protocols (Newell & Simon, 1972; Lewis, 1982; Lewis & Mack, 1982) could lend more insight to user errors, to the users' cognitive model of the system, and to the alternative user strategies which are indicative of mismatching the users' cognitive model. Another point for future testing would be to better classify user errors and relate then to the users' cognitive model, in a similar manner to Norman (1982). The use of cluster analysis techniques for classifying errors similar to Folley and Williges (1982) could lead to a better understanding of the users' cognitive model. Experimental comparisons of various features of the interface can provide evidence for interface design guidelines. Potential areas to be studied include an alternative to the hierarchically designed menu structure, a different dialog design, and a comparison between the menu driven interface and the System/34 interface.

CONCLUSION

Several benefits resulted from this research. First, the prototype or simulation of the interface allowed the design team to use the interface to get hands-on experience, which in turn permitted obvious design flaws to be corrected. Demonstrations of the interface were made to gain managerial support for the research project. The simulation was critical for the human factors evaluation that resulted.

There were several benefits that resulted from the evaluation methodology. By being iterative in nature, the evaluation allowed the interface to be redesigned and subsequently reevaluated to determine the success of the changes. Research involving user–computer interface design almost necessarily needs to be iterative. The probability analysis was successful in identifying weak points in the hierarchical menu structure.

The error analysis provided a better understanding about the types of user errors in using menus. Finally, the evaluation methodology pointed to aspects of the users' cognitive model of the system and validated the success of this approach for a menu driven user interface.

The interface itself has several benefits. It is a menu driven interface which easily encompasses the novice/casual user. There are multiple levels of the interface (e.g., keywords) to avoid penalizing the experienced/sophisticated user. The interface is comprehensive in that all System/34 functions are accessible. The evaluation has demonstrated that little prior knowledge and no offline assistance (e.g., manuals) are needed to successfully use the interface. Flexibility was designed into the interface so the user never gets trapped; there is always a way to exit a menu or a prompt. The evaluation has shown that the hierarchical design of the menu driven user interface has either matched or complemented the users' cognitive model of the system. Redundant paths allow for alternative user strategies or mismatched user models. Finally, extensive use of help text will be involved in the next phase of the design and evaluation of the interface.

The basic concept of this menu driven user interface is clearly sound. Comments received from participants were generally favorable and performance reasonably successful. Any future changes made to the interface will be more "fine tuning" than "major overhaul." Finally, extensive research is desperately needed to better understand how to design user–computer interfaces to match users' cognitive models and perceptions about computer systems.

ACKNOWLEDGMENTS

Many people were involved in the design and evaluation of this interface and in reviewing this paper. Those who contributed significantly were: Nancy Blackstad, Steve Dahl, Karen Eikenhorst, Dave Peterson, Mike Temple, and Jim Abraham. Special thanks goes to Thomas Barnhart for his developing and programming of the prototype and to his overall involvement with the design of this interface. Finally, we would like to acknowledge Ms. Judy Knoke for the preparation of this manuscript.

REFERENCES

Brosey, M. K., & Shneiderman, B. Two experimental comparisons of relational and hierarchical data base models. *International Journal of Man–Machine Studies*, 1978, *10*, 625–637.

Dray, S. M., Ogden, W. G., & Vestewig, R. E. Measuring performance with a menu-selection human-computer interface. *Proceedings of the 25th annual meeting of the human factors society*, Rochester, NY, 1981.

Durding, B. M., Becker, C. A., & Gould, J. D. Data organization. *Human Factors*, 1977, *19*, 1, 1–14.

Engel, S. E., & Granda, R. E. Guidelines for man/display interfaces. IBM Poughkeepsie Laboratory Technical Report TR00.2720, December 19, 1975.

Folley, L., & Williges, R. User models of text editing command languages. *Proceedings of the human factors in computer systems*, Gaithersburg, MD, 1982.

Lewis, C. Using the "thinking-aloud" method in cognitive interface design, IBM Research Report RC 9265 (#40713), IBM Thomas J. Watson Research Center, Yorktown Heights, NY, 1982.

Lewis, C., & Mack, R., Learning to use a text processing system: Evidence from "thinking-aloud" protocol, *Proceedings of the human factors in computer systems*, Gaithersburg, MD, 1982.

Martin, J. *Design of man–computer dialogues*. Englewood Cliffs, NJ: Prentice Hall, 1973.

Newell, A. and Simon, H. A., *Human problem solving*. Englewood Cliffs, NJ: Prentice Hall, 1972.

Norman, D. Steps forward a cognitive engineering: Design rules based on analysis of human error. *Proceedings of the human factors in computer systems*, Gaithersburg, MD, 1982.

Peterson, D. E. Screen design guidelines. *Small Systems World*, February, 1979, 19–21, 34–37.

Ramsey, H. R., & Atwood, M. E. Human factors in computer systems: a review of the literature. Englewood, CO: Science Application, Inc., Technical Report SAI-79-111-DEN, September 1979.

Shneiderman, B. *Software psychology*. Cambridge, MA: Winthrop Publishers, 1980.

8

Statistical Semantics:
Analysis of the Potential Performance
of Keyword Information Systems*

G. W. FURNAS
T. K. LANDAUER
L. M. GOMEZ
S. T. DUMAIS

This paper examines how imprecision in the way humans name things might limit how well a computer can guess what they are referring to. People were asked to name things in a variety of domains: instructions for text-editing operations, index words for cooking recipes, categories for "want ads," and descriptions of common objects. We found that random pairs of people used the same word for an object only 10–20% of the time. But we also found that hit rates could be increased threefold by using norms on naming to pick optimal names, by recognizing as many of the users' various words as possible, and by allowing the user and the system several guesses in trying to hit upon the desired target.

INTRODUCTION

Computer-based information management systems can store, manipulate, and transmit enormous quantities of information. They can allow almost unlimited organization, multiple indexing, and cross referencing, and they are capable of performing rapid and complex search operations. Thus, they can provide far more powerful tools for knowledge management than have previously been available. Such tools will be important to almost everybody: cooks wanting to find recipes, doctors needing patients' histories, managers tracking inventories, clerks filling orders and keeping records, and buyers looking for products. It would be fine if such systems could be used directly by inexperienced and occasional users, people whose main jobs or talents lie elsewhere. The hope is that new information-seeking tools could be operated with no

*Reprinted with permission from *The Bell System Technical Journal.* Copyright 1983, AT&T.

more training than is needed to use a card file or a book index, but with much greater success.

We believe that some of the greatest difficulties blocking this dream are psychological. Certainly, more available speed and capacity and new algorithmic and data structure developments involving deep and difficult problems will be needed before the arrival of fully satisfying information systems. But, progress along these lines has already been enormous and continues at a brisk pace.

Meanwhile, while even existing computational capabilities are very great, our ability to make them easily usable by nonspecialists has been quite limited. An important psychological problem is in understanding the relationship between what people say and what they want. This understanding is the key to designing systems that can infer what services or information users need from the input they provide. Information systems of many sorts need this basic capability; bibliographic reference search systems, business management databases, airline reservation systems, plant inventory and customer record systems, even text processors. We believe that current systems generally fall very far short of the ideal of always knowing just what to give the user.

At this time, although it is clear that there is a problem, very little relevant work has been done (see Carroll, 1982; Landauer, Galotti, & Hartwell, in press). We do not even know the locus of the problem in the chain of actors and events. For example, the main problems may lie at the source; perhaps the human intellect is basically incapable of forming information specifications that are very precise. If so, perhaps no system could do much better than current ones. But, it is also possible to believe that if systems could understand people as well as, say, close friends understand each other, there would be fewer problems.

We have tried to get some idea of the extent of this problem by attacking a small but important part of it. We have asked people to give descriptions of various information objects, and analyzed their responses in order to determine how well the objects to which they refer can be inferred from what they say. We have begun by studying the referential properties of isolated words and short phrases.

Our goals in this research have been twofold; first, to advance understanding of the psychological processes by which human semantic reference is generated, and second, to model and estimate the strengths and weaknesses of information systems that take human-generated descriptions of sought items as their input. Empirical observations of naming behavior provide the necessary data for both enterprises.

In this chapter we first describe four sets of object–description data, the way they were collected and reduced, and some of their more interesting features. Then we present a series of analyses in which we treat the collected descriptions both as representative of what users would provide as input to information retrieval systems and as the source of the information that the system would use in determining user wants. The hypothetical systems we consider are limited to ones in which the user's initial entry or query consists of a single word or short phrase. We do not attempt to analyze the possible performance of systems that make use of sentential syntax or linguistic or real world context. The reason for this limitation is largely pragmatic; it postpones analysis of many difficult complexities. However, the characteristics and limitations of single-word to object reference which we have investigated have strong implications for many access methods (by which we mean the access method provided for the user and not that used by the program).

We are, of course, aware that there are data access methods that do not start with the user entering a key word or phrase; for example, there are strictly menu-driven systems, ones that rely on well-formed queries and restricted query languages, and also ones that attempt some form of natural language understanding. These other methods share some, but not all, of the same conceptual and practical difficulties of key word methods, and each raises somewhat unique and interesting psychological issues of its own. Where our data bear on these issues in reasonably direct ways, we offer some comments, but we focus primarily on the key word comprehension process and its ramifications. In the final section of this analysis, we discuss some of the reasons why key word access is as limited as we find it to be and consider several potential methods for overcoming the deficiencies.

DESCRIPTION OF DATA SETS

There are things a computer system has or does that a user might wish to refer to. These "information-objects" are just the objects, e.g., operations or data sets, to which commands and queries apply. We have collected descriptions of information objects in four quite disparate domains, chosen in an intentional effort to achieve variety, and also for their relevance to a number of special problems that are outside the concerns of this study. The four sets are: the verbs used in spontaneous descriptions of the operations needed to perform manual text-editing operations, descriptions of named common objects designed to induce another person or a computer to return the name in a passwordlike game, superordinate category names for items available in a swap-and-sale listing similar to classified ads in newspapers, and index words provided for a set of main-course cooking recipes.

In this section we will describe how the object specifications were obtained from people, how they were reduced to single word or short-phrase keys, and summarize a few qualitative and statistical characteristics of the responses.

Text-editing Operations

The first data came from language applied to text editing. The study was one of two conducted to explore "naturalness" in command names (Landauer et al., 1983). Forty-eight secretarial and high school students were asked to provide instructions to another hypothetical typist describing what operations needed to be performed on a text marked by an author for correction. These corrections involved two examples each of 25 sorts of edits: five basic operations (insert, delete, move, change, and transpose) on each of five textual units (blanks, characters, words, lines, and paragraphs).

Preprocessing in this case reduced each response to the main verb or phrase in the instruction describing how to perform the editing operation. While this was in fact accomplished manually here, we believe that a simple parser and English word list could, in principle, have given nearly identical results. These expressions may be considered candidates for command names for editing systems.

Perhaps the most striking result was that there was extensive disagreement in the verbs people produced. This point is the main focus of the current article and will be dealt with in considerable detail later. For now, let us just make a few preliminary notes, e.g., that the three most popular names for each operation accounted for only 33% of the total number of responses. The intersubject agreement, the probability that any two people used the same verb in describing a particular text correction, is only .08. Since each of the 25 sorts of edits occurred twice, we also have a measure of within-subject (with 1200 observations) agreement. The probability that an individual subject used the same main verb in the two cases was .34.

What agreement there was did not favor the terminology used by our locally popular editor (the UNIX,[1] Bell Laboratories' text editor *ed*.) For 24 out of 25 of the types of edits, the name in *ed* was not the most frequent spontaneously given name. Use of the terms "delete" and "substitute" was quite rare, for example. (Landauer et al., 1983, went on to show, however, that this caused no problems in initial learning on the basic editor.) People preferred "add" for the insert operations, "omit" for delete operations, and "change" for the replace operations. There was little consensus in describing the transpose and move operations.

[1]UNIX is a trademark of Bell Telephone Laboratories, Incorporated.

Common Object Descriptions

These data were originally collected in a study by Dumais and Landauer (unpublished), examining how people naturally obtain information from one another. The information objects here were names of 50 common items chosen from 10 "categories": cities, proper names, clothing, animals, food, household items, abstract words, a category of words with highly associated opposites (e.g., black, love), and two categories whose members were words selected in such a way that negation might figure strongly in the descriptions, e.g., to eliminate the unwanted set members. Three hundred thirty-seven New York University students were asked to write down a description that would enable another person (or in half the cases, a hypothetical computer) to guess the object. There were no restrictions as to the form or content of the descriptions, except that they could not contain the target word itself. Subjects also indicated whether or not they had any computer experience. Those with at least one computer course were classified as computer "experienced" in the data summaries discussed below.

A subsequent study was conducted to evaluate the effectiveness of the descriptions generated in the first study. Twenty-five subjects (6 Murray Hill area homemakers and 19 Bell Telephone Laboratories employees) were each given 150 descriptions randomly selected from those generated by the NYU students. They were asked to: (a) guess (without knowing the alternative) the item being described, and (b) indicate on a 5-point scale their confidence in their guess.

Principal results from the main study include the finding that when communications were intended for computers, people with computer experience were relatively more terse, and nonexperienced people were relatively more verbose than when communications were intended for people. However, there was no simple relationship between verbosity and effectiveness, i.e., guessing accuracy as indicated by the second study. People were somewhat more successful in guessing the target items when the descriptions had been provided by people without computer experience (81.2% vs. 78.5%), but this difference fails to reach statistical significance.

A point of considerable interest here was the style of specification that subjects used. Subjects' descriptions were not very precise, typically they refer to a whole set of items, not just the intended target (although we have no good measure of this other than informal ratings of denotative class size). Still, the average successful guess rate of the second group of subjects was over 80%. We will return to a discussion of this paradoxically high success rate toward the end of this chapter. The most frequent way of specifying target items, used about 60% of the time, was to describe them in terms of a superordinate (sometimes followed by characteristics or attributes which distinguish the intended target from other members of the superordinate category). Another fairly common form of description (~20%) was the use of exemplars. For several of the target words (e.g., motorcycle, magazine, sports, games, science), subjects listed examples for more specific items falling into the target category (e.g., Harley, Suzuki . . . , in the case of motorcycle), instead of attempting a more formal definition. Negations (and opposites) were used less than 50% of the time for the words we thought were particularly amenable to this form of description.

For the purposes of the analyses undertaken in the current study, these descriptions were preprocessed to merge minor variations in as automatic a fashion as possible: uppercase was folded to lowercase, word endings were stripped (plurals, tense markers, etc.), and "noncontent" words (including articles, imperatives, conjunctions, prepositions, pronouns, and tenses of the verb "to be") were removed.

An average of 8 words per description were in this way condensed to an average of 5.4 words. The first of the remaining words (i.e., first standardized content word) was tabulated for the statistical analyses.

Superordinate Categories for Swap-and-Sale Items

The major purpose of collecting these data was to develop empirical networks of "ISA" relations, that is, classification hierarchies based on user knowledge and representations, for a set of items to be incorporated in an experimental menu-driven information access system (Furnas, unpublished).

The information objects were 64 items taken randomly from roughly 300 entries on a monthly bulletin board listing of items for swap and sale at Bell Laboratories, NJ. The subjects were 30 local New Jersey homemakers. Each subject worked with a random 32 of the 64 target items. They were told that they were participating in the study to find out how they classified various items being sold on local bulletin boards. The use of these categories in helping people in future computerized retrieval systems was mentioned. Subjects were instructed to complete successive "All _____ are _____ "sentences. Beginning with the specific target they were to give its immediate superordinate (e.g., "all *red delicious apples for sale @10¢ ea.* are *apples*"). Then they copied the first given superordinate ("apples") to the beginning of the next incomplete sentence and finished the new sentence with a still more general category (e.g., "all *apples* are *fruit*"). They were to continue in this way (e.g., "all *fruits* are *food*") until they could go no further. They were then to go back and find some category that had another superordinate in addition to the one they had already cited, and list that category with its new superordinate. These data were also standardized by stripping off endings and discarding noncontent words.

On average, an individual subject produced 2.1 different chains of successively more general superordinate categories for each stimulus. The chains averaged 2.5 superordinates each, with superordinate categories named in phrases containing an average of 1.7 standardized (content) words. For the current analysis, only the lowest level (most specific) category from the first generated chain of superordinates was used. The category name was used in its (standardized) entirety.

The categories given in this study make it apparent that the construction of a network that will faithfully match all users' conceptions of a domain is not an easy matter. People have difficulty in generating superordinates, and show considerable disagreement as to how things should be grouped under those superordinates. Categorization and indexing schemes currently in use always depend on a user either generating the same superordinate as the system knows about, or at least being able to choose the right one from a list. Perhaps the difficulty and lack of agreement among people in categorizing information objects accounts for much of the perceived deficiency of current menu-driven data access methods.

Recipe Index Words

The original motive for this data set was to study the effect of domain expertise (i.e., cooking skill) on indexing and keyword usage (Gomez & Kraut, unpublished). The information objects were 188 main course cooking recipes (French, Italian, Mexican, and American cuisine) taken from 12 cookbooks of explicitly varied sophistication (ranging from a garden club's cookbook (*Our Favorite Recipes: Inverness Garden Club*, 1977–78), through *The New York Times Cookbook* (Claiborne, 1961), to *The Art of French Cooking* (Child, Bertholle, & Beck, 1979).

There were three groups of 8 subjects each: experts, who taught cooking classes; and intermediates and novices, selected from local homemakers who came out at the high and low extremes of several self-rating scales on culinary sophistication.

Subjects were told their task was to describe each of the recipes in keyword form, selecting at least three but no more than seven descriptive words or brief phrases for each recipe. They were told that their job was similar to that of a librarian who is creating an index or card catalogue, and that the descriptions should be useful to another person trying to locate that recipe in a large set of recipes. Half of the subjects in each experience group were instructed to direct their descriptions to expert cooks using the index, the other half to novice cooks. The description tasks was self-paced by each subject in her home. Subjects required between 5 and 10 hours to complete the task.

Terms here were again preprocessed to remove word endings and noncontent words. All multiple-word productions were scored by two judges (the experimenters) to determine if the phrase could be decomposed into its constituent words and maintain its meaning. Subjects produced an average of 5.4 key words per recipe, the first, "most important" of which was studied here.

Again, we found considerable diversity. For the 188 recipes, a total of 303 different word types were used by the 8 experts, 220 by the 8 intermediates, and 252 by the 8 novices. It is interesting to speculate on the reasons why these groups differ. Perhaps the experts have a large and specialized vocabulary and the novices have an unruly, haphazard one. In any case, there seems to be something more conventional about the word use of intermediates, a point to which we will return later.

General Comments on the Data Sets

These data sets all pertain to information objects that one might want accessible on a computer. They were also all of modest size. Other than that, though, they tapped very different knowledge domains, they asked for specification in a number of different ways, and they were provided by different kinds of people. Moreover, the method of reducing the free-form descriptions given by our participants to single words and short phrases varied somewhat from one case to the other. This variety of data is important to our purposes. In order for results to have any pretense of robustness, it is important that they be obtained on a sufficient variety of cases to assure that it is not the particulars of the objects at hand that are responsible for the observed characteristics. We know of no way to actually sample data domains, descriptive methods, and reduction methods in a representative way. However, we believe that results which hold for all of the disparate sets that we have studied stand a good chance of holding for most others.

For each domain, our data can be represented as a table in which the rows are words provided by the subject, the columns are the objects to which these words were applied as descriptions, and the cell entries indicate the number of times each word was used in the description of a given information object. The questions we ask concern how the information contained in such a table might be used to guess from an input word what object is intended (see Tables 1 and 2). In fact, there is an implicit third dimension to these tables representing the person from whom the description was obtained, and sometimes a fourth represent-

TABLE 1
Sample Data from the Text-Editing Study[a]

Words	Objects				
	Insert	Delete	Replace	Move	Transpose
Change	30	22	60	30	41
Remove	0	21	12	17	5
Spell	4	14	13	12	10
Reverse	0	0	0	0	27
Leave	10	0	0	1	0
Make Into	0	4	0	0	1
. . .					

[a]Numbers are the frequency with which a word is used to refer to an object.

ing which of several words of a multiple-word description provided by a given subject is involved. However, for most of the analyses we consider only the first word given and the matrices are all very sparse, so we have chosen to collapse across subjects. It is worth noting that the tables are not sparse simply because we have failed to collect enough data. Word usage tends to resemble Zipf's distribution (Zipf, 1949) in that a few words are used very frequently and many words used only once (see Figure 1). As more and more data are collected some cells increase in frequency, but the number of unique words also grows so that the sparseness of the table tends to be preserved. Moreover, most words refer only to a limited number of objects, so that such tables usually have a large number of empty cells.

TABLE 2
Sample Data from the Common Object Study[a]

Words	Objects					
	Calculator	Lime	Lucille Ball	Pear	Raisin	Robin . . .
Small	17	0	0	0	7	4
Machine	4	0	0	0	0	0
Green	0	18	0	7	0	0
Bird	0	0	0	0	0	21
Fruit	0	1	0	19	1	0
Red	0	0	8	0	0	7
Female	0	0	2	0	0	0
. . .						

[a]Numbers are the frequency with which a word is used to refer to an object.

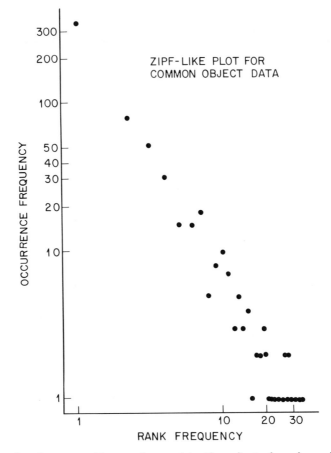

FIGURE 1. **Plot of common object word usage data. The ordinate shows log rank frequency and the abscissa shows log occurrence frequency. Many words (over 340 here) occur only once, and few words occur very frequently. Zipf's law is the supposedly straight line relation between occurrence frequency and rank frequency when both are plotted logarithmically.**

INTRODUCTION TO ANALYSES TO DETERMINE
REFERENT EFFECTIVENESS

In the analyses that we report in the next section, we have been interested in how much information about referent object identity is contained in the words used to describe them. In order to consider how good a word or expression is as a reference to some object, it is necessary to suppose some sort of a mechanism by which the expression is comprehended. We can only estimate the value of inputs as determinants of output if we can specify the function that takes one into the other. The inputs of our problem are the words provided by people in specifying information objects. The output is a guess or set of guesses about which object was intended. Our approach has been to specify various ways in which the input can be made to yield the output. We have considered models or functions that are intended to mimic what happens in typical information systems and also ones that try to improve on current methods in various ways. We also develop models that allow us to estimate what ideal input–output mappings could accomplish. It is in this sense of ideal or asymptotic performance that we can appraise the limits of word reference.

The models that we examine have in common that they use the frequency table of users' word reference, and only this data, to guide the system in guessing users' intents.

The General Model

In the coming sections we consider a number of special cases of the following rather general baseline model. In order to retrieve a desired item, the user is assumed to enter one (though in some instances, K different) key word(s). The system is assumed to make either one or M guesses as to the user's intent. Success is counted if the intended object is among the system guesses.

As an actual system, our baseline model would operate as follows. The user would enter one (or more) key words or phrases. If it could recognize the input at all, the system would return one or more potential objects according to some rule, perhaps relying on data in a table of observed word-use frequencies. (If it could not recognize the input, it would return nothing.) In cases where the system returns several alternative objects, it can be assumed that the user would be able to select from among them the one intended without error. This corresponds to a simple case of what might be called a "menu-on-the-fly" technique, in which the first entry is a freely chosen key word, and the next step is a choice among a menu of items selected on the basis of the system's knowledge about referents the human user might have intended by that key word.

We will now describe the models that we have analyzed, beginning with the simplest ones, those closest to the way typical systems are currently set up, and proceeding to more sophisticated models. Each model is characterized by a set of assumptions or constraints on the input-output function of the system.

The following sections, consider a number of models and analyses in considerable technical detail. Some readers may wish to skip these sections and pick up at the section entitled "Highlights."

To give a preview, the results convince us of three major points. First, the vocabulary problem is a serious one for untutored users of computers, particularly given the poor naming techniques typical of current practice. Second, the difficulties come from a severe and fundamental lack of consensus in the language community on what to call things. Finally, our research suggests that, by using a very nonstandard approach, namely trying to make the best possible guesses on all user words, substantial improvements could be made in meeting the underlying objective of providing access to the things people want.

Formalizing the General Model

Let us make things more formal. We have two sorts of entities, the user-generated inputs, referred to here as "words," "terms," or "descriptors" and the system outputs, called "information objects," "targets," or just "objects." We are concerned with two relationships between these entities. On the system's side are the associations dictating how the system will respond to a user's words; which words will be associated with what system objects. On the user's side are the actual intended associations; what the user in fact meant by a given word. Both the system's and the user's relationships can be shown in the form of a table. The *system table* gives the input–output mapping of which object the system will retrieve (execute, etc.) for what words, with a 1 wherever a word is associated with a system object, and a 0 otherwise. The *user table* shows the frequencies with which people say what words for what objects. Examples of a user table were shown in Tables 1 and 2. These tables can used in the evaluation, and, if one wishes, in the design of the system table.

The System Table

The system table has two aspects. First is what might be called the set of structural constraints. These concern how much and what variety of input the system will respond to, and are imposed by capacity limitations on memory, processing, data collection, or sometimes by the computer algorithms used. Second is the choice of a particular instance of the input–output system table. That is, the assignment of exactly which words the system will respond to, and with what objects. One design problem centers on whether assignments will be made in an effective or possibly optimal way.

Structural constraints in the system's input–output table correspond to general restrictions on which cells may be used in the mapping of input to output. We consider patterns based on limiting the number of cells in rows and columns (e.g., the numbers of words understood as referring to each object and the number of objects taken as possible referents for each word). In particular, we will be considering models where, in the system table, there are various restrictions on

1. total words recognized by the system (number of nonzero rows in the system table);

2. number of words recognized for each object (column totals);
3. total objects referenced (number of nonzero columns);
4. number of objects referenced by each word (row totals); and
5. total word–object associations "understood" by the system (grand total of cells).

Once a general structure has been determined, one still must choose which associations to give the system, i.e., exactly which cells to include in the mapping. We consider three cases which give rise to different versions of each model. The first is a purely random assignment. Within the specified structural constraints, the cell defining the system's associations between words and objects are chosen randomly. Clearly, this is not a realistic system model, but it is useful in the conceptual analysis of other models and as a baseline of comparison. The second case is a weighted random assignment, assigning a word to an object with the same probability that the word and object occur together in usage. That is, input–output associations are chosen with probabilities proportional to their frequency of use, as indicated by the user table. This is, in fact, a very important case to consider, since it approximates the way many systems are currently designed. We will often refer to this as the "armchair" method: A single human designer sits in his or her proverbial armchair and makes the name–object associations by some sort of intuitive guessing. The user table is, in effect, a compendium of many humans' "armchair" attempts to assign good words to the system objects. A weighted random sample from the table thus corresponds to a random person's "armchair" nomination. Such weighted sampling therefore allows us to estimate the effectiveness of "armchair" design. The last assignment technique is by the best available optimization procedure. Optimization methods make generous use of the empirical data in the user table to pick the best cells. For some system constraints we do not know a combinatorially feasible way to find the best configuration, but it is possible to improve substantially upon the weighted random method. Note that the success of any optimization attempt is also limited by the quality of the data available in the user table.

The User Table

The user table is a data matrix compiled from the studies metioned earlier in this chapter. It records the frequency with which untrained users employed various terms in referring to the system objects.

We assume that the descriptions given by our subjects for the information objects in the various sets bear a close resemblance to the entries similar users would make in trying to specify the same objects in a database system. This assumption may be wrong in detail; in using actual systems people might give descriptions somewhat different than those induced by our instructions. However, when we varied the intended recipient of the descriptions, as in the recipe (experts, novices) and common object data (people, machines), there were only small changes in the descriptions. So we believe that descriptive language would change little in attempts to communicate with real, as compared to hypothetical, systems. (We cannot, however, estimate the effect that prolonged interaction with a given system might have on a user's vocabulary.)

We also assume that collection of user descriptions provides a good basis for predicting actual user intent from key word input. In approaching a data access system, the user must have some information object in mind, and must give a description to the system. In actual use, it is difficult to know what the user really has in mind. Sometimes the user has no clear idea at all. It is not apparent whether this "diffuse target" case is easier or harder for a system to deal with. A user with an unclear goal may or may not be more easily satisfied, but is probably less likely to give precise specifications.

It is obvious, however, that the diffuse target case is harder to study. Reliably inducing such a state in the user is difficult, and the system's success is not easily appraised. Therefore we have collected data and investigated models based on the clear target case.

Thus, in these models it seems to us that the best information the system could start with is to know what typical users would give as descriptions, given the intended object was well-known and clearly defined. Our assumption is that the descriptions people give of objects when we specify them will closely resemble the description they would give if they had thought of the specific objects themselves. This seemed a practicable and, we hope, realistic starting point.

Going Through the Models

Insofar as possible, we will use a consistent expository framework to describe each model. The framework is as follows:

1. *Name:* A mnemonic title, by which the model will be referenced.
2. *Interpretation:* What the model is all about.
3. *Motivations:* Where and why it might come up.
4. *Structural Constraints:* The general form of the system's input–output table ("system table") for the model. This will be given as clearly as possible in English, with a more formal summary of the constraints for each model appearing in Appendix A. This appendix is a very useful reference source in comparing the models.
5. *Analyses:* For each of three methods of selecting cells for the system table (RANDOM, WEIGHTED RANDOM, and OPTIMIZED), the following information will be given whenever possible.

a. *version:* what the method means in this model.
b. *evaluation by:* what statistic is used to assess its success.
c. *result:* Statistics are computed for the four data sets:
  (1a) Edit command data, all 25 Operation × Text-Unit combinations (abbreviated: "Ed25")
  (1b) For purposes of comparison we include a second analysis of the same data: the 5 basic Operations collapsed across textual units ("Ed5")
  (2) Common Object data, all 50 objects ("CmOb")
  (3) Swap-and-Sale data, the 64 sale items ("Swap")
  (4) Recipe data, the 188 main-course recipes ("Recp")
d. The following statistics will be given:
  (1) Recall prob: how often the method succeeds in returning the desired object.
  (2) Mean number returned: when the system is able to make a guess about the user's desire object, this is a record of how many guesses it makes to achieve the reported probability of recall.

A few comments are needed about the "number returned" statistic. First, note its importance in considering recall success probabilities. If the system returns a large number of guesses, it can obviously be expected to have a greater chance of including the target. Thus, comparisons of performance demand that this number be kept in mind. Second, note that these indicate the number of things returned when the system in fact recognizes the user's input and so is in fact able to hazard a guess. For some systems it will be common to have insufficient data and to make no guess at all for many user words.

We denote the number of objects, or columns, in the table by C, and the number of user words, or rows, by R. The former is in part well-defined by the designers of the system, but the latter, the vocabulary size, is usually an artifact of data collection, as it would tend to expand if more data were collected. Here it refers to the size of the vocabulary in the corpus of data we collected.

ANALYSES TO DETERMINE REFERENT EFFECTIVENESS

Model 1: One Name per Object

Interpretation

Each object in the system is assigned a single term or name. The user enters one term. Success depends on the user's word coinciding with the system's.

Motivations

This approach is common in computer systems—each entity has one and only one name. The name is usually chosen by the designer or by an expert indexer. The designer hopes to establish a convention about what system objects will be called. Users must either learn or guess the names to make the system work. Such learning may be feasible in small systems or for highly practiced users. The growing community of novice and infrequent users, however, are often reduced to trying to second-guess the system, even to find documentation. This is often frustrating and we shall see that in principle the approach is far from adequate. Moreover, the real difficulties for such a scheme often go unremarked simply because traditional computer users have come to accept as normal the necessity of extended learning, repeated second-guessing, lengthy searches or expert consultation in finding the correct names for programming commands, file names, or information categories.

Structure

The only constraint is that there be exactly one word for each object. Note especially that this model allows the possibility that word can be used to name more than one object. This is a situation that designers often try to avoid. It nevertheless arises in several situations, as when programs, commands, file names, or index categories are assigned by many independent users, or to highly similar objects (like bibliographic subject or author specifications, or the case of several functions being collapsed under a single command name). We evaluate models that disallow such "collisions" later. Note that for the likelihood of recalling a given object at all, the aspect of the problem on which we focus in these first analyses, assigning a word to more than one object as we do here, can only improve the expected system performance.

Analyses

RANDOM Version. For each object one name is chosen randomly from the total vocabulary for that object, i.e., one cell from each column. That is, the name of each object is a random choice among all the total set of descriptors given to all the objects.

evaluation by: theoretical value

In this case we can calculate the expected performance of the system exactly. The success rate is given simply by the ratio of t, the total number of cells included in the system mapping, to RC, the total number of cells in the matrix. A simple proof of this appears in Appendix C.

result:

recall prob = C/RC = $1/R$, i.e., the reciprocal of the total number of words that users
        use for all the objects

             Ed5 Ed25 CmOb Swap Recp
recall prob = .012 .008 .002 .001 .001
mean number returned = $1 + (C-1)/R$

The "mean number returned" is a measure of the amount of ambiguity or imprecision in the terms as the system understands them. It is the average number of things the system knows by a given name that the system recognizes; the system must return all of these objects when trying to guess a target. For some of our models the number of objects that the system returns will be fixed ahead of time by design, but here it is the mean of a random variable, easily calculated to be $1 + (C-1)/R$.[2]

Weighted Random Version. A system name is assigned to an object with the same probability that users attributed the term to the object. Here the user enters one key word; and the system has had a key word assigned to each object, based on one other person's nomination. This model mimics, more or less, a currently fairly typical situation for key-word systems (or program-name or command-name access systems) in which the system designer has provided an entry name for each object, obtained only from one usage datum (the designer or indexer's "armchair" introspection), and the user is required to enter just that name. We assume that system designers or indexers are like our subjects in their choice of names, so that the relative frequency with which a name was given to an object by our subjects provides an estimate of the likelihood that a designer would choose that name for that object. (One source of support for this assumption is that experts gave no more consistent names than novices; see below.)

                evaluation by: the "column repeat rate" statistic

We estimated how well such a system is likely to work by estimating the probability that a given word chosen randomly from our population

---

[2]Under the pure random method, each object has one of the R words associated with it randomly and independently. Thus for each object, any given word arising or not becomes a Bernoulli event with probability $1/R$. So having chosen a word for one object, the number of the other $C-1$ objects having randomly been assigned that same word is given by a binomial distribution with parameters $p = 1/R$ and $C-1$. Thus the mean number of other objects with the same name as our given object is the mean of this distribution $(C-1/R$. The system will return any of these objects plus the original object, or $1+(C-1)/R$ objects.

(e.g., by one user on one occasion) would match another word chosen from the same population (e.g., by one designer). This probability is known as the repeat rate (Herdan, 1960). If we index rows (words) by i, and columns (objects) by j, in a word-by-object table, then repeat rate for a given object, rep_j is defined as:

$$r_j = \sum_i Pi_j^2$$

This formula can be understood by considering that a match between two randomly chosen words occurs when, for any given word first chosen the second word is the same. Say the first word was $word_i$, the second word will match it with the probability of $word_i$ occurring in the population, p_{ij}. The probability of $word_i$ being the first word is also p_{ij}, so the probability of $word_i$ being involved in a match is p_{ij}. Summing across all possible words, we get the equation given above.

An unbiased estimate of the population probability of such a match comes from calculating the true probability of such a coincidence in drawing from our sample without replacement, given by:

$$\hat{r}_j = \sum_i \frac{n_{ij}}{N} \frac{n_{ij} - 1}{N - 1}$$

where N is a column total and n_{ij} is the frequency of the ith cell in the column j.

Here we are interested in the average probability of success, given any particular target, throughout the table. So we decompose the overall probability by conditionalizing on columns, calculating the repeat rate for each column, and then average these success probabilities, using weights proportional to the column total frequencies. (In our data these column totals are approximately equal by design.) These weighted average column repeat rates are given below for our four sets of data. These numbers may be interpreted as answering the following question: Given that all objects are equally often the desired target, with what probability would the name be given by a user trying to specify a target match the name assigned to it by a designer?

result:

	Ed5	Ed25	CmOb	Swap	Recp
recall prob =	.07	.11	.12	.14	.18

While there is some variation among the values, they are all quite small. People do not agree with one another very well as to the first word or phrase with which to label an object. The probability that two typists will use the same main verb in describing an edit operation is less that 1 in 15. The probability that two people will use the same first key word for a recipe is less than one in five. (These numbers also tell us something about the size of the set of alternatives that people use in their disagreement. It is a property of repeat rate that the set of alternatives must be at least $1/r$, and can be quite a bit larger if they are not equally likely, as is the case here.)

Most of the interesting comparisons will be between the different models presented (e.g., this model with the subsequent ones), and not the different data sets. To facilitate model to model comparison, all results appear together in a summary table in Appendix B.[3] The reader is strongly encouraged to refer to this table, throughout this section.

The mimicked method in which the designer provides the system with only one entry word that it can understand, and the user enters just one key word, is clearly unsatisfactory for untrained users. The usual solution has been for system designers to rely on users learning the chosen vocabulary, i.e., to try to force the user's table to adapt to a fairly arbitrary system table. When the system is small and the user's interaction frequent, this can work quite well. Indeed, Landauer et al. (1983) have shown that using unrelated random names has little or no detrimental effect on intially learning to operate a small editor. But, if the system is large and its use intermittent and nonexpert (as for example in large-scale information retrieval systems like library catalogues, recipe files, or classified product catalogues), it is simply unreasonable to require users to learn a specialized vocabulary.[4] Despite the designers intentions, the uninitiated will try to make the system work without memorizing extensive naming conventions. Thus the problem remains a real one.

One approach we might consider at this point is to seek expert advice in choosing names. This is a fairly common approach, taken in the hope that experts in a given subject area know what things are, or should be called and so might generate words for more general currency. Indeed, the indexes to books, libraries, user manuals, and other information sources are customarily created either by subject matter experts or by professional indexers.

---

[3]Of course some comparisons between data sets are also of interest. Note for example that the value for Ed25 is higher than for Ed5. This says that the set of words applied to the individual objects is more sharply restricted than for the collapsed classes. This is to be expected, since any diversity between objects in the pattern of terms applied becomes within-class diversity, when the objects are collapsed together, driving the column repeat rate down. Only if all objects in a class had identical naming patterns would the repeat rate not decrease.

[4]In intermediate cases, like program and command names for an operating system, the method may be satisfactory for expert users, while leading to dissatisfaction for others (see Norman, 1981; Lesk, 1981).

TABLE 3
Repeat Rate Measures of Agreement Between Indexers and Users for the Recipe Data Set[a]

	ee	en	ie	in	ne	nn
ee	.11	.17	.14	.15	.10	.22
en	.17	.11	.18	.17	.16	.21
ie	.14	.18	.20	.20	.21	.23
in	.15	.17	.20	.19	.17	.26
ne	.10	.16	.21	.17	.16	.16
nn	.22	.21	.23	.26	.16	.32

[a]The three groups of key word providers (experts, intermediates, and novices) are subdivided by whether they produced key words for experts or novices. Thus, the "ne" group is a group of novice cooks who produced key words intended for use by experts.

We have collected some data relevant to this issue in the recipe study. Our key word providers represented several levels of expertise. The situation is not unrepresentative; usually the indexes of cookbooks and recipe collections are created by cooking experts, presumably on the assumption that their characterization and labeling will be superior, even for less sophisticated users. We calculated repeat rates separately for the three groups of keyword providers (experts, intermediates, and novices), subdivided by whether they had produced the key words under instructions to make them appropriate to novices or to experts. The results are shown in Table 3. The repeat rates shown were calculated in a special way. For each cell, the index words were provided by particular subsets of describers, and the proportion of matches was calculated on a pool of descriptor words provided by other subjects. Thus, the (ee, ee) cell estimates how likely a word provided by one expert for other experts was to match that provided by another expert with the same, expert, audience in mind. The (en, ne) cell estimates the probability that a word provided by an expert for novice users would match the word provided by a novice for expert users. Clearly the differences among marginal values (i.e., the averages of repeat rates for different index providers) are not large, and more clearly still, expert cooks do not provide better descriptions for the use of either other experts or novices. (If anything, novices do the best in using one another's words.)

The usual armchair approach, even if undertaken by subject matter experts, has only a small rate of success. The obvious step at this point is to seek explicitly optimal choices of names, treating this as the empirical question that it clearly is.

OPTIMIZED (Best) Version. If we want to use the name that has the greatest currency among subjects, we must choose the term that is in fact

maximally used by subjects for each object. We pick the maximum cell in each column and use the corresponding term.

> *evaluation by:* various estimates of the range of expected performance (a lower bound based on split-halves analysis; and upper bounds from a transformation of the column repeat rate and from an analysis that assumes probabilities)

To know the performance of this model we need to know the true population magnitude of the maximum cell in a column. That cell is the one we would choose according to an optimum name assignment scheme, and its size would be the proportion of future users' terms for the given object that would coincide with our optimal choice. Problems arise in that there are no known distribution-free, unbiased estimators of the population magnitude of the maximum probability cell. We have, however, been able to devise a few techniques that let us put bounds on the performance of this system. The first uses a split-halves technique to give a lower-bound estimate. The data in each column is split into two halves, and the maximum cell chosen on the basis of the first half (as though we had to design our "optimal" system on the basis of half the data). This cell is then matched against the second half, to see how well it succeeds. Thus the second half acts as a virtual experimental test of the performance of the "optimal" method. The split half results shown here are the average performance of ten independent splits of each data set.

> *result:*
>
> | | Ed5 | Ed25 | CmOb | Swap | Recp |
> |--------------|-----|------|------|------|------|
> | recall prob = | .15 | .19 | .26 | .26 | .31 |

These numbers are an unbiased estimate of how well a system would do, using this optimum strategy but constrained to a small amount of data (namely the amount in each half). It clearly underestimates how well one could do with more data. We have used two approaches to obtaining an upper bound. One is based on an interesting inequality relation that holds between the size of the maximum cell and the repeat rate statistic described above: It can be shown that the former is no greater than the square root of the latter.[5]

---

[5]This relation follows from two inequalities. First, note that the repeat rate is the sum of the squared cell probabilities. This sum is clearly greater than or equal to any of its terms, since all are positive. In particular, it is larger than the square of the maximum probability cell. Thus, the expectation of the repeat rate is larger than the expectation of the squared maximum cell. Second, note that the expectation of any squared variable is always greater than or equal to the square of that variable's expectation. Putting these together, the expectation of the repeat rate is greater than or equal to the square of the expected magnitude of the maximum cell. Thus an upper bound on the maximum cell is estimated by the square root of the repeat rate.

Thus, the performance of the optimum model is expected to be no greater than the square root of the performance of the weighted random (armchair) model. This statistic will overestimate performance to the degree that individual words other than the maximal one are also applied frequently to a given object. Since our own data suggest that this is commonly true, this upper bound is likely to be quite generous. It has an important pragmatic advantage, however, in that it is independent of sample size and easy to obtain. Other estimates of optimal performance require collecting detailed data on the precise pattern of naming. This estimate requires only that one observe the probability that two people use the same name (i.e., the repeat rate), without even having to note what particular terms they use. Taking the square root then yields an upper bound estimate for the best possible single name per object. For our data, these quantities are:

result:

	Ed5	Ed25	CmOb	Swap	Recp
recall prob =	.27	.32	.33	.35	.42

The final way to estimate the performance is to let the data predict itself. The observed largest proportion falling in a cell is taken as the estimate of the population maximum. A familiar result in rank-order statistics is that in the presence of error the observed maximum of a number of observations is expected to be larger than it should be. Thus using the maximum to estimate itself is likely to be an overestimate for any limited sample size. This is particularly true for small samples. In the extreme case, note that if a column of cells has only one observation, the observed maximum will be in the one cell where the single observation fell, which will have an estimated probability of 1, regardless of the true underlying probabilities. This will become more relevant later when we calculate similar statistics on rows of the matrix, where many of them will involve small numbers of total observations in each row. For now, however, the samples are not too small, and the results are presented below.

result:

	Ed5	Ed25	CmOb	Swap	Recp
recall prob =	.16	.22	.28	.34	.36

Discussion

It appears that performance would be about twice as good using an optimum naming strategy than for the weighted random model (which we believe to be an approximation to typical current practice) in a one-word per object system. Still, the overall levels are not very impressive. It should be noted that these optimal strategies represent the best *any* single-name scheme could do. No expert, human or otherwise, could choose single names that would work better.

Is this the best that we can hope to do for people? The answer is no. There are other, more effective approaches. One is to use multiple names, sometimes called "aliases," for each object, which leads us to our second general structure for a system model.

Model 2: Several Names per Object

Interpretation

Each object in the system is assigned M terms or names. The user enters one term. Success depends on the user's word coinciding with any one of the system's M words.

Motivations

Giving things single names is not adequate for the uninitiated, so a reasonable next step is to try giving the system several names by which to recognize each object.

Structure

The constraint is that exactly M words are stored for each object.

Analyses

RANDOM Version. For each object M names are chosen randomly from the total vocabulary, i.e., M cells are chosen from each column. The recall probability is M/R, and in our sample tables would range from .01 to .04 for three names per object ($M = 3$).

WEIGHTED RANDOM Version. The first name is chosen with a probability proportional to its frequency in the given object's column of the table. Each successive name is then similarly chosen, without replacement, from the remaining cells. This is as though someone "sitting in an armchair" thought up several distinct names, any one of which the user could use with success. We make an independence assumption, that the probability of a word being chosen is independent, except for renormalization, of the words already chosen. This is probably a faulty approximation for a single human generating a series of words, where the first word thought of may influence the next. If the words were separately proposed by different designers, this independence assumption might be more nearly correct.

Note that by reversing the roles of the system and the user we obtain a very interesting dual interpretation of this model. That is, let the system store a single word and the user make M different guesses. Success would be counted if any of the user's M words matches the system word. Either of these interpretations merely involves the probability that a single sample from the column will match one of M samples drawn without replacement from the cells of the same column (that is, without replacing the whole cell, not just the observation from the cell).

evaluation by: "M-order repeat rate" statistic, within columns

These success probabilities can be estimated using an extension of the the repeat rate statistic to this case of multiple samplings without re-placement of types. The probability for column j can be shown to be estimated by:

$$\hat{r}_j^{[m]} = \sum_{i_1} \frac{n_{i_1 j}}{N} \frac{n_{i_1 j} - 1}{N - 1} \left[1 + \sum_{i_2 \neq i_1} \frac{n_{i_2 j}}{N - n_{i_1 j} - 1} \left[1 + \sum_{i_3 \neq i_1 i_2} \frac{n_{i_3 j}}{N - n_{i_2 j} - n_{i_1 j} - 1} \left[1 + \cdots + \sum_{i_m \neq i_1 i_2 \cdots i_{m-1}} \frac{n_{i_m j}}{N - \left(\sum_{k=2}^{m-1} n_{i_k j} \right) - 1} \left[1 \right] \cdots \right] \right] \right]$$

This formula requires a calculation time that grows exponentially in M. We therefore limit results to $M \leq 3$.

result:

	Ed5	Ed25	CmOb	Swap	Recp
recall prob					
($M = 1$) =	.07	.11	.12	.14	.18
($M = 2$) =	.14	.21	.21	.25	.33
($M = 3$) =	.21	.30	.28	.34	.45

Note that allowing the system (or equivalently the user) several words for each object in trying to match its partner achieves considerable gain. Performance almost doubles and triples as we go to two and three guesses.

It should be remembered that the estimated probabilities of success for 1, 2, 3 guesses were calculated on data from subjects first responses only. Under the interpretation of this model in which subjects make repeated guesses at a system's single word, this is equivalent to assuming that successive guesses, given that previous guesses had failed, would resemble other first-provided words. To get a true estimate of the probable performance of a system that actually used this technique, we would need data on people actually making successive guesses. One would have to know how many such guesses can actually be made and with what quality before knowing what the maximum performance would be. However, supposing that the system, or perhaps the user's desires and abilities, limited input to three guesses, performance of such a system would not be likely to exceed about one chance in three of correct return.

Optimized VERSION. Choose as the M names for each object those terms that are maximally used by subjects for the object; that is, pick the highest M cell in each column and use the corresponding terms. This is a simple generalization of the single name case.

evaluation by: split halves; self-predict

We again run into the problems involved in estimating maximum, and now also the nearly maximum, cells. The split-half and self-estimation procedures were used here to estimate upper and lower bound. The split half results represent the average performance of ten independent splits of each data set. The results are presented below:

result:

	Ed5	Ed25	CmOb	Swap	Recp	
recall prob						
(M = 1) .15	.19	.26	.26	.31	(split half)
.16	.22	.28	.34	.36	(self-estim.)
(M = 2) .26	.32	.36	.36	.49	(split half)
.28	.37	.41	.50	.56	(self-estim.)
(M = 3) .37	.42	.42	.45	.58	(split half)
.38	.49	.48	.59	.67	(self-estim.)

Discussion

Considerable benefit is obtained both from using data to optimize choices and from giving the system several names for each object (or equivalently allowing the user several guesses). Though it was not undertaken here, it would be interesting to explore the possibility of letting both the user and the system use several names to try to match each other.

The improvement gained by storing multiple words has associated with it a potential cost in ambiguity. Nowhere in either this model or the previous, one-name model has there been any concern that the names be distinct. The names for each object were picked from the corresponding column, with no regard for what names were being chosen for other objects. This means that two objects could be assigned the same name, with consequent ambiguity should the user give that name. The system would be unable to tell which object the user intended, and would have to present the user with a menu of choices to differentiate among them. (Recall that at the outset we stated that we would be assuming such a system, and, moreover, that choices among the items on the subsequent "menu-on-the-fly" would be assumed error free.)

Naming choices will collide when two different objects have the same words applied to them. One might expect this to happen, for example, if two objects are very similar, so that the same words apply to them. We have collected some data that confirm the existence of this similarity effect. For both the Recipe and the Common Object data we compared the probability of a naming collision for random and similar objects in the set. Subjects were given 5 "focal" objects and 25 "match" objects, all drawn at random from the set. For each of the focal objects, they were told to select the one match object that was most similar to it. Using the weighted random naming method, the naming collision rates were compared for these five pairs of similar objects and the same five focal objects paired with random members of the match set. That is, we calculated the probability that a word applied by one user to the first object would be the same as the word applied by another user to a second object. Using subject differences as a random error estimate, there was a highly significant increase in naming collisions for the pairs of objects judged to be similar ($t(14) = 3.86$ and $t(24) = 5.37$ for the Recipe and Common Object data, respectively).

To examine the effects of trying to avoid collisions, we studied the next model.

Model 3: "A Distinct Name for Each Object"

Interpretation

Each object in the system is assigned a single term, with the proviso that no term may be used more than once. Success depends on the user's spontaneously produced word coinciding with the system's distinct name for the intended object.

Motivations

Note that in typical naming circumstances, the intent is often to establish conventions about terminology so as to avoid the ambiguity that would otherwise arise. Naming conventions are used not only to set by fiat the name by which an object will be known to a system, but also to proclaim that *only* that object will be so known. Often interaction will reach a stage where complete precision is needed, e.g., for actual command execution.

It may be the designer's motivation in selecting names, that users learn terminology which allows them to be precise. As has already been pointed out, however, the designers intent may not correspond with the user's reality; the untutored may try to deal with the system anyway. Thus we explore the user-guess success for systems designed with the unique name constraint.

The models discussed here are just like those of the "one name" case, except that words cannot be used more than once.

Structure

One distinct word is stored for each object, i.e., this is the same as Model 1, except that the words must all be distinct, so that at most one object is referenced by each word. This imposes strong constraints on the system table; on both row and column totals, and the number of rows and columns used overall. This high degree of constraint makes mathematical treatment difficult, as will be elaborated shortly.

Analyses

RANDOM Version. For each object, one name is chosen randomly from the total vocabulary. Once a vocabulary item is used, however, it is eliminated from any future consideration.

Recall probability would be $1/R$ and range from .002 to .014 for these data. The "number returned" is easy to give for this case, as it is set by the structural constraint, that each name have a unique referent. It is therefore for all models having this structure. That is, when the system finds a target it returns exactly one candidate. However, there are many occasions when the user word matches no system word and so no target is returned.

WEIGHTED RANDOM Version. Here the name is chosen with a probability proportional to its frequency in the given object's column of the table. Then the chosen word is eliminated, and a name is chosen in an analogous way for another object, from the remaining words, etc. The results are clearly influenced by the sequence in which objects are dealt with. The approach used here was to choose in the following way: A word/object pair is chosen by a weighted random sampling from the whole table. This gives the right distribution within each column and allows appropriate representation of each column. Once such a cell is chosen, the corresponding object has been named and the word used, so both are eliminated, and the procedure is iterated on the table, now reduced by one row and column, until all objects have been named.

evaluation by: monte carlo simulations with split halves

We could devise no way to evaluate this model analytically, so we used a Monte Carlo simulation on split halves. We divided the data in the user table in half and randomly picked names, according to the model outlined above, using the data from one half. The second half of the data was then used to evaluate the effectiveness of the names thus chosen. The results presented here are the average for ten split halves with ten independent Monte Carlo simulations of name selection in each.

result:

	Ed5	Ed25	CmOb	Swap	Recp
recall prob =	.07	.08	.11	.12	.09
number returned =	1.0 (by design)				

Note that, as might be expected, success is less than for the comparable model in which names did not have to be distinct. In some instances this decrease is small, as with the edit studies, in others it is large, as with the recipes. This should depend on whether there were a few high frequency words that were used for many different objects.

Analysis of Precision versus Popularity of Terms. In all of our data sets there was a slight trend for high frequency words to be less discriminating than low frequency words. In studying this relation quantitatively, the measure of discriminating power of a word was given by the repeat rate statistic, this time applied within each row. This row repeat rate is a measure of the probability that two uses of a word refer to the same object. When this probability is high, the word is very discriminating. Below we present the correlation (Spearman r) of row repeat rates with the marginal frequency of the words.

correlations:

	Ed5	Ed25	CmOb	Swap	Recp
corr:	−.28	−.30	−.21	−.24	−.16
(N words):	(74)	(84)	(301)	(145)	(167)

This correlation points to the cause of the difficulties run into by the constraint of distinct names. There is something of a conflict. If one chooses high-frequency names, they will be likely to collide. If one chooses less frequent names, the chances of collision will be somewhat less, but fewer user's queries will be handled. Unfortunately, there is no correspondence between the size of these correlations and the size of the decrease in performance for each data set. The reasons for this are not clear.

OPTIMIZED (Though Not Best Possible) Version. The idea is to choose the distinct names such that the total probability in the cells chosen is maximal. The method used here is a greedy algorithm that is not truly optimal, but it is a reasonable improvement over the weighted random method. It begins by picking the highest cell in the matrix, then after eliminating the corresponding row and column from the matrix, it iterates, picking the next highest cell, etc., until all objects have been assigned a name. Algorithms like this are called "greedy," because at each step they take the biggest possible chunk of what is left, without regard for what later problems that may cause. In this case, it is possible that an early choice will eliminate a word that would be very good for another object, when there was an alternate word that would have done almost as well at no such cost. Algorithms which would take such future complications into consideration, or equivalently consider so many possibilities at once, are typically combinatorially explosive. Thus we present the results of the straightforward greedy algorithm approach.

evaluation by: split halves

There is no simple method to estimate the performance of such an approach. Again we turned to the split halves technique of dividing the data in half, applying the algorithm to one half and then testing the chosen names on the other half. The results, averaged over ten independent splits, are given below.

result:

	Ed5	Ed25	CmOb	Swap	Recp
recall prob =	.14	.11	.23	.19	.11
number returned = 1.0 (by design)					

Discussion

As expected again, the optimization attempt, even though imperfect, has a dramatic effect, in many cases doubling the performance of the system. Even in the best case, however, the system succeeds only about one quarter of the time, and typically little more than a tenth of the time. We note that the numbers here are substantially lower than in the case where there was no requirement that names be distinct. In fact this improved method really does no better than an armchair (weighted random) version of the unconstrained model, and often worse. The lesson here is that adding the requirement of uniqueness, common in establishing conventions, hurts naive users quite a bit. The need for such conventions is not denied, but it is simply asserted that auxiliary aids will be needed for systems that make heavy use of such conventions. It is worth noting that for our example data sets the number of objects that are not disambiguated and would have to be presented for a second stage choice is always small, and the gain in recall always large. Thus, the "menu-on-the-fly" method implicit in several of the models presented here appears to be very promising as an aid to unsophisticated users.

Model 4: Distinct Names, Augmented with M Extra Referents

Interpretation

Each object is given a distinct name; $M-1$ other referents for those words are also stored.

Motivations

Part of the inferiority of the distinct name models, when compared to the unconstrained models, came from the fact that people often want to use the same names to refer to several objects. The motivation for the next model is to recapture access to some of those other interpretations of the term. A simple extension of the distinct name structure is to begin with the situation where each object is associated with a distinct name, one that can be memorized and used unambiguously by experts, but also to store $M-1$ other objects that the term applies to, explicitly for use in a "menu-on-the-fly" for the untutored. A reasonable system implementation might be to give the "distinct" name a special status (the "real" meaning of the term), and to admit the other interpretations as secondary, perhaps to be verified in the context of use.

This is the first model in which we explictly design in more than a single system guess as to the intended object. To be sure, several guesses for the meaning of a term could (and would) have arisen in the uncontrolled name cases of Models 1 and 2, but here we predetermine exactly how many guesses the system makes and evaluate performance as a function of this parameter.

As more objects are returned, the likelihood of including the intended target is increased, but a certain cost is incurred, the cost associated with discerning the true target from amongst all the returned objects. Data searches can fail not only by not giving access to a desired object, but also by returning too many unwanted objects. In information science the problem is familiar as the tradeoff between *recall*—the number of desirable items returned—and *precision*—the proportion of items returned that are desirable. This is also similar to the hit versus false alarm tradeoffs of signal detection, familiar to psychologists and communication theorists. In the latter context, the tradeoff is sometimes examined by tracing out operating characteristics and interpreting them in terms of parameters of an unerlying statistical theory. There does not currently exist a corresponding statistical theory for our precision and recall rates, but it is useful to be able to examine, or in this case control explicitly, precision and see what gains in recall result. This yields an operating characteristic. Any version of the general baseline model in which the system is set up to return an explicit number of guesses provides a direct way to do this.

Structure

At least one name is given to each object, but each name that is used by the system also refers to M different system objects.

Analyses

RANDOM Version. Here the primary cells are again chosen completely blindly, eliminating rows and columns already used in an iterative manner exactly like the distinct name case. After the C primary cells are chosen, the $M-1$ additional or secondary cells are chosen randomly from each row chosen.

The resulting probability is M/R and would range from .001 to .035. The number returned is, of course, M by design for all versions of Model 4.

WEIGHTED RANDOM Version. Cells are chosen at each step with a probability reflecting their magnitudes. Thus first C distinct names are picked, as in the distinct name model (Model 3). Then, in the row of each cell just picked, $M-1$ more objects are chosen with a probability equal to their relative frequencies in the rows.

Again, the weighted random case is an approximation to what an individual designer might do, without collecting data from other people. The process of coming up with additional interpretations requires a bit of further explanation. The asymmetries found in free association data suggest that people starting from terms and generating objects would yield probabilities quite different from those obtained from people starting with objects and generating terms (Keppel & Strand, 1970). This asymmetry means, in this model, that the additional interpretations cannot be thought of as the result of the designer sitting in an armschair and thinking up other interpretations for the terms. The efficacy of such an approach is not estimable from our data.

A scenario that might better satisfy the assumptions of this model would have the system memorize the first $M-1$ nonstandard uses of its known terms that it comes across. The number of guesses that the system is required to achieve this performance is of course, M, the number returned by design.

evaluation by: Monte Carlo simulation on split halves

    result:

	Ed5	Ed25	CmOb	Swap	Recp
recall prob =					
(M = 1)	.07	.08	.11	.13	.08
(M = 2)	.13	.15	.15	.17	.15
(M = 3)	.18	.21	.18	.20	.20

OPTIMIZED (Though Not Best Possible) Version. Again it was not feasible to find the completely optimal choice of cells. Instead, the primary cells were found in the same "greedy" way used for the simple distinct name "optimal" case. The subsequent choice of secondary referents for the words so selected can be optimized by choosing the $M-1$ largest cells remaining in the row. Note that it is possible for some of these remaining cells to be larger than the primary cell, as when they refer to an object that was eliminated before the row was chosen. (It is the vagaries of the need for convention in names that brings about this ironic use of the term "primary.")

evaluated by: split halves

In the now familiar procedure, the data are split and the first half used to choose the cells following the algorithm just outlined. Then the second half of the data is used to evaluate the choice of cells.

    result:

	Ed5	Ed25	CmOb	Swap	Recp
recall prob =					
(M = 1)	.14	.11	.22	.21	.11
(M = 2)	.21	.20	.28	.27	.17
(M = 3)	.27	.29	.31	.30	.22
number returned = M (by design)					

These results, for one to three total guesses, are plotted in Figure 2. The model curves can be viewed as operating characteristics, giving total recall as a function of decreasing precision. The dashed diagonal lines represent expected chance performance if the system returns guesses at random. Note that in the case of the Ed5 data, the random performance is *better* than that of this model. This is possible because this model, like most current computer systems, makes no response at all to any but a few words. By failing on so many of the words it encounters, it can be surpassed by a system that only guesses, but at least guesses for all words. Admittedly, in the context of command execution any level of guessing without confirmation may be dangerously out of place, but there are safer arenas, like help facilities, which could benefit. In any case the cost of ignoring so many user words is made clear.

Notice that the total probability does not approach one, even in the

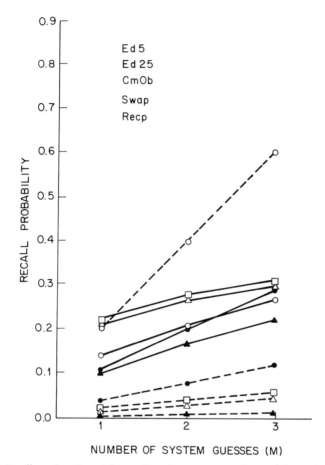

FIGURE 2. **Recall as a function of the number of system guesses for Model 4 ("Distinct Names, Augmented with *M* Extra Referents") are shown by the solid lines. The dashed lines represent expected chance performance if the system returns guesses at random. (Different chance performance for the different data sets simply reflects the different number of objects in each set.)**

case of command verbs for editing operations where there were only five objects. The reason is again that many words were used by our subjects to describe these objects, and the algorithm picked only a maximum of *C* of them to recognize. The modeled assumption was that if a subject supplied any word not recognized by the system, no choices would be returned and a failure would result.

Discussion

These models use the same number of cells as the multiple names models, yet do not do as well. The implication, presumably, is that there is considerable cost in limiting oneself to a small number of words. The users distribute their descriptors too broadly for any such limitation to be very satisfactory.

Thus we are motivated to try a completely different approach. All the models up to this point have inherently focused on what the system brings to the interaction. The modus operandi has been to set all the system's objects before us and then try various ways to guess what users will call them, albeit with some improvements as we give deference to the empirical characteristics of the user's language as seen in the user table.

Suppose we turn the tables, so to speak, and focus on what the users bring to the interaction: the terms they use when trying to specify things. Start with the terms, all of them, and concentrate on trying to guess to what object they refer. This brings us to our next model.

Model 5: Recognize One Referent for Every Word

Interpretation

The system stores every word that it can, and for each has one guess as to the intended object.

Motivations

One of the principal problems with the approaches studied so far has been that a large proportion of the recall failures can be attributed simply to not recognizing many of the user's words. The systems modeled have paid too little attention to the wide variety of the terms people use spontaneously. Here we focus on this essential character of what the users bring to the interaction. For every word produced during the collection of the data for the user table, a guess is made. The different versions vary as to how these guesses are made.

It is useful to note that these models and their generalizations in the next section are really duals of the single and multiple names models. In those models the system had words for each object, and it succeeded when its name was the same as the word the user would use. Thus to evaluate those models, we calculated the conditional probability that, given a target object, the system and user names would coincide. Then we summed probabilities across objects, weighting each by its column's marginal probability (all essentially equal in our data, by design). Here we conditionalize the other way. Given the word used, we find the probabilities that the system and user associate the same object with the word. Then we sum across words (rows), weighting by the very uneven observed marginal frequencies for each word, to get the overall performance.

Structure

The system makes use of every word in the table, associating each with one, and only one, system object.

Analyses

RANDOM Version. Here the system will associate random objects with the input words. That is, the system will simply make a random guess for anything the user says. The recall probability is just $1/C$, and ranges from .005 for the recipes to .200 for the Ed5 data. The number of guesses being made to achieve these performances is predetermined by design to be 1 for all these models.

WEIGHTED RANDOM Version.   Here an object is associated with each word by a random method that takes each object with probability proportional to its relative frequency in the row. Arguments similar to those made for the augmented distinct name model here imply that this is not the analog of having the designer sit down with the list of words and hazard guesses as to what they mean; the data used here go in the other direction. The more appropriate scenario is as if a system records the first encounter with a new word, and its intended referent, and makes a single pointer based on this single observation, forever freezing the meaning of the word. This is perhaps an unreasonable scenario, but the model is worth investigating because the statistics that result have other valuable interpretations.

evaluation by: row-repeat rate statistic

The probability of success in this case is the probability that two users will mean the same thing by a given word. The results below are the average row repeat rates, weighted by total row frequency. (Rows with a frequency of one were excluded, since the repeat rate is not calculable in that case.) This number is an unbiased estimate of the overall probability that any two occurrences of any word will be in reference to the same object.

result:

	Ed5	Ed25	CmOb	Swap	Recp
recall prob =	.41	.15	.52	.62	.13
number returned = 1.0 (by design)					

We note how variable the probabilities are. They are low where the same words mean different things, and high where any one word refers to only one object. Apparently these domains differ in this aspect. There are at least two possible explanations. First is the similarity effect discussed in connection with Model 3. There it was shown that pairs of objects judged to be similar had higher overlap in the patterns of names applied to them than random pairs. The resulting naming collisions meant that it was more difficult to find distinct names for similar pairs. Here we are explicitly interested in a different, though related effect. We can use the previous experimental data (see section on "Analysis of Precision versus Popularity of Terms") again, this time to demonstrate the similarity effect on a row-repeat rate measure. In this version we consider pairs of cells in each row: one cell from the target column, and one from either a column rated similar or from a random control column. The repeat rate is calculated on the two cells and summed (weighted by row sum) down all the rows. Since a low repeat rate for a given word indicates that it does not discriminate well between the objects, we predict a lower repeat rate for the similar pairs of objects, and that is what we obtain. In the Recipe data the mean repeat rate of the similar pairs was .73 and for the random pairs was .91 ($t(14) = -5.83$). For the Common Object data, the similar pair repeat rate was .89 and the random repeat rate was .95 ($t(24) = -5.71$). Thus similarity has a strong effect on repeat rate. Unfortunately, what we need to make sense of the varied results is some measure of the relative internal similarities of the various domains. Such data is neither available nor easily obtained. Still, it might be agreed that the set of Common Objects is more diverse than is a set of Cooking Recipes, or Text Editor operations, corresponding to observed differences in average repeat rates.

There is another factor which might be helping the Swap-and-Sale descriptors. Analysis of the other data sets was either strictly limited to single words, or else only to short phrases with few content words. Swap-and-Sale descriptors contained an average of 1.7 content words, where there were fewer than 1.2 content words in the others. If these words tend to be used in a conjunctive sense and if they are not redundant, they should contribute to the high selectivity of the descriptors seen for the Swap-and-Sale data.

A final note should be made of the higher repeat rate for Ed5 compared to Ed25. Part of this is due simply to the fact that even with a purely random, undiscriminating distribution of name usage across objects there would be an increase in repeat rate as the number of objects is decreased. If there are only two objects, people will have to mean the same object by any given term at least half the time. A more intriguing possibility is that the classes are more distinct entities than are the individual objects, and that this makes word usage more discriminating, above and beyond the chance effect just mentioned.

OPTIMIZED Version.   In this case the user matrix is used to find the best possible choice for a word's referent. This is done by picking the maximum cell from each row, the object to which the word has most commonly referred.

Evaluation by: split halves;
square root of the row-repeat rate;
self-prediction

Evaluation is again problematic. While there is no question that picking the maximum cell is the best possible strategy, there is no unbiased estimate of its true magnitude. Thus we resort to the same three methods used when similarly confronted in the optimal version of the one-name model (Model 1): the split half technique; and unbiased estimate of how well one could expect to do with half the data; and two estimates of upper bounds.

result:

	Ed5	Ed25	CmOb	Swap	Recp	
recall prob =						
	.49	.18	.43	.35	.13	(split half)
	.62	.35	.65	.72	.28	(sqrt of row rep rate)
	.54	.26	.69	.81	.25	(self predict)

number returned = 1.0 (by design)

Discussion

There is considerable variability, both in the performance and in the range of these estimates. The largest range is for the Swap-and-Sale estimates. This is due to the very large number of descriptors that occurred only a few times. The data for each such descriptor is statistically unreliable, so the maximum cell cannot be accurately identified, a problem severely aggravated by splitting the data and only using half to make the identifications. Note that this is not an experimental artifact. A very long tail on a descriptor distribution is a legitimate real-word problem, because it means that new terms will keep arising that the system will not have seen before, and will thus will be able to make no educated guesses about. A wide range in the estimate reflects the fact that very large amounts of data would be needed to approach asymptotic performance.

As for the overall diversity of scores from domain to domain, the arguments about similarity and number of words in the descriptors, given above for the repeat rate case, are equally applicable here.

The most important point to be made about this optimized model is that it represents the best one could possibly do. We can do no better than to recognize every word users try and make an optimal guess as to its meaning. Clearly, if we have insufficient data to do this well, either preventing us from recognizing a word, or from being able to estimate the modal referent, performance will suffer. But the upper-bound estimates represent the real and strict limits of performance. No other pattern of structural constraints could do better, except at the cost of precision. This tradeoff with precision is explored in our final model; in which an explicitly prespecified number of multiple guesses is returned to increase the chance of including the user's intended object among them.

Model 6: *M* Referents for Every Word

Interpretation

The system stores every word that it can, together with a set of M possible referents. Whenever the user enters a word, the system returns the M guesses, possibly ranked in some way.

Motivations

The "one referent for every word" model (Model 5), in its optimized version, gave the best possible performance for a system that hazards only one guess for a user's word. The only way to increase the chance of returning the user's intended object is to return more than a single guess. These guesses could be returned in the form of a menu of M items, among which users would choose.

As mentioned, this is the dual of the multiple name set of models, and several of the evaluation procedures differ only in that rows have traded roles with columns.

Structure

Like the previous model, the system makes use of every word in the table, but here it associates M system objects with each.

Analyses

RANDOM Version. The system makes just M pure guesses, without replacement, from the set of system objects. The recall probability for these cases is thus exactly M/C, and the number returned is M, by design, for all versions of Model 6.

WEIGHTED RANDOM Version. In a manner following the single referent, weighted random model, the M cells are chosen with a probability that is proportional to their relative sizes. The M choices are required to be distinct, and so a cell is excluded from further consideration once it has been selected.

The scenario that this might correspond to would be one in which the system learns the first M distinct referents of the word that it comes

across in use. Its encounters would be governed by exactly these probabilities.

evaluation by: M-order repeat rate statistic, within rows

We want the probability that the user's single intended referent coin-
cides with any of the system's candidates, when both sets are drawn from
the same probability matrix. These success probabilities are estimated
using the same extension of the repeat rate statistic used in evaluating
success of the many names for one object model (Model 2b, weighted
random version). Here though, it is applied to the cells within a row,
rather than within a column.

It should be noted that the formula requires there to be M nonempty
cells in the row to prevent some of the denominators from going to zero.
For example, it is not possible to give an unbiased estimate on how well
three guesses would do if the data only show two objects. Note that while
rows with highly discriminating words should in general have high M-
order repeat rates, it is exactly such rows that will be less likely to satisfy
this requirement, especially at lower marginal frequencies. Thus, ignoring rows where this statistic cannot be calculated biases the average value
of this statistic downward. Though it is not clear how to get an upper
estimate, there is another way to get a lower bound, by recognizing that
performance for M guesses is at least as good as performance for $M-1$
guesses. Thus a lower-bound on the average performance comes from
using for each row the calculable repeat rate of highest-order not greater than M.

result:

	Ed5	Ed25	CmOb	Swap	Recp
recall prob =					
($M = 1$)	.41	.15	.52	.62	.13
($M = 2$)	.66	.26	.65	.74	.21
($M = 3$)	.81	.36	.71	.80	.27

OPTIMIZED Version. This version takes the optimal strategy for multiple guesses. The user gives one word, and the system makes guesses that
the user means one of the M most likely things the word has meant in the
past, as deduced from the data in the user table. That is, it returns the
objects associated with the M highest cells in the row corresponding to
the word given. (A sequential version of the model would give the user
the guesses in decreasing observed frequency, beginning with the maximum cell.)

evaluation by: self-prediction;
split halves;

As in other cases where evaluation of performance depends on the estimation of true population ordered frequencies, we must resort to indirect methods to give a range. We use split halves to give a lower bound and self-estimate to give an upper bound.

result:

	Ed5	Ed25	CmOb	Swap	Recp	
recall prob =						
(M = 1)	.49	.18	.43	.35	.13	(split half)
	.54	.26	.69	.81	.25	(self predict)
(M = 2)	.70	.30	.53	.43	.20	(split half)
	.76	.42	.82	.92	.37	(self predict)
(M = 3)	.83	.40	.58	.47	.26	(split half)
	.88	.53	.88	.96	.46	(self predict)
number returned = M (by design)

The recall rates, as estimated by the conservative split-half procedure are given in the Figure 3. The curves can be interpreted as operating characteristics; the distance between the curves and the corresponding dashed diagonal line reflects how much the implied system could be expected to outperform a simple menu system based on the pure random model given above, i.e., on a random choice of M objects. All of our models do substantially better than chance. More importantly, they do much better than the augmented distinct name model (Model 4) presented in Figure 2.

Discussion

There are really two important points to be made about these results. The first is that these performances are variable and but moderately high. Note that the very high value of the three system guess case in Ed5 is rather vacuous, since there were only 5 objects to guess from. In the Common Object case, however, the high values are quite meaningful, since there were 50 objects.

The legitimate high values lead us to the second observation, that substantial improvement has been gained over the traditional, one-distinct-name approach. The presumption is that this gain is due in part to the increase in capturing all the user's inputs. There are, after all, two ways for the system to fail in retrieving the user's target. One is by making wrong guesses, the other is by being unable to make any guesses at all. To give an idea of the size of this problem: for three of our data sets, no system using only C words (e.g., one for each object) could recognize even half the user's words. Thus, recall rates can only increase as the system is taught more words, and often this can be a sizable improvement. The only conceivable decrements (except those associated with time and space in index management) would be if the additional

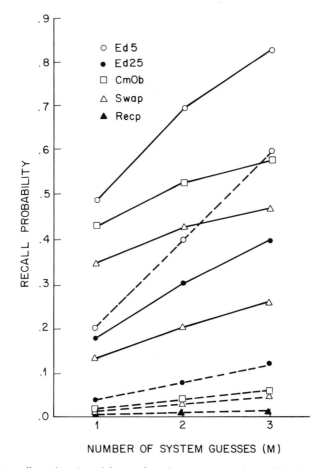

FIGURE 3. **Recall as a function of the number of system guesses for Model 6 ("M Referents for Every Word") are shown by the solid lines (split half estimation). The dashed lines represent expected chance performance if the system returns guesses at random. (Different chance performance for the different data sets simply reflects the different number of objects in each set.)**

words were less precise, thus decreasing precision for a fixed recall rate, or decreasing recall for a fixed precision. But our finding of a consistent, albeit small, negative correlation between a word's frequency and its selectivity suggest that the opposite is true. As a greater number of lower frequency words are included, precision will go up.

Recall also increases dramatically if the choice of algorithms is optimized, though never exceeding the square root of the performance obtained by "armchairing," i.e., making pointers based on a single observation. The cost is in the trouble it may take to collect data. In some cases the variety of terms is so great that data collection must continue for quite a time before asymptotic performance is approached. Performance also increases as the system is allowed to make more guesses; this involves a direct trade of recall probability for reduced precision. In many circumstances, particularly in help facilities, this presents little problem. The user need only be given the various options, and whatever additional information it takes to decide what is really sought.

Other Possible Models and Some Limits

The models presented so far have covered a large share of the space of simply constrained models where the user makes one guess, but they do not exhaust the ways in which a system could use empirical data on user descriptions to guess the user's desired objects.

One might consider, for instance, models that take advantage of redundancy in the tables. A good theory or description of what such tables are like could be used to augment the data to make better estimates of true population usage probabilities. For one example, if we knew that some single shape of distribution characterized all columns, we could use the data to estimate parameters of that distribution instead of individual cell probabilities. This would reduce the number of parameters that need to be estimated from the data and consequently improve stability and reliability of predictions.

A related, and perhaps more interesting, idea would be to look for latent structure in the similarity between objects, and between words, and use this to improve predictions. For example, if we knew or could infer from the data that two words were used almost identically, we might pool the cell frequencies for the two words. Or if we knew that the objects and words related to each other by some more elaborate structure (e.g., a hierarchical tree) we could base predictions on calculated indirect reference paths.

While these ideas might be useful, the research presented here puts strict limits on the success of any such approaches. As noted above in Models 5 and 6, the self-prediction evaluation procedure (determining the best guesses from the data and then using the same data to estimate success) and the "square root of repeat rate" evaluation procedure yield upper bounds on expected performance. Indeed, no analysis method of the kind we have just been discussing, no matter how clever, could improve on this result. The reason is as follows. Even using inherent structure to improve predictions could do no more than increase the accuracy of estimation of the population values for the input–output tables. Even with true population probabilities, however, the guessing decision rule would be the same; choose first the maximumly probable referent of each word, and so forth. The input–output tables are inherently probabilistic as a result of disagreement in word usage, not just of the estimation uncertainties. For a given user population, with a given degree of training, a certain amount of disagreement in word usage will occur. Even with perfect knowledge of the probabilities describing this usage, we could not predict individual referents perfectly. Now, the objective of having true population probabilities as cell entries is mimicked by our upper-bound estimating procedure. It pretends that the values observed for each cell are exactly the probabilities that would exist in further sampling from the population. Thus, no method of "cooking" the data to reduce sampling error could achieve a better result than is illustrated by our upper-bound values.

To improve performance beyond these limits, it would be necessary to construct systems that use different input, e.g., multiple words or interactive dialogues, or make the user learn to use more easily interpretable language. The last of these is the current default approach—make the user learn everything—and has its limits for large systems or occasional users. Thus we believe a more fruitful direction to explore further is that of multiple word inputs.

The simplest models of multiple word query might assume only content words (no syntax) and a statistical independence between words. Independence would allow the performance of such a system to be estimated from single word data of the sort we have collected. Multiple words could be used disjunctively. In this case the system would return any objects that matched any of the input words. This would result in an increase in probability of recall. Alternately, the words could be used conjunctively, so that the system would return only objects matching all the keywords. The result would be higher precision. Mixtures of these two approaches (e.g., conjuncts of disjuncts) could clearly then be used to improve both precision and recall.

Of course, from our data we do not know how people can or do use multiple words, even without considering syntax. The popular balances of conjunction and disjunction are unknown, and spontaneous multiple guesses are no doubt not independent. But what sort of nonindependence is common? For example, do multiple terms tend to be more unrelated or more redundant than independence predicts? These questions must be explored before multiple-word systems can be theoretically evaluated or optimized for the naive user.

The consideration of syntax leads to a whole new set of problems, those often encountered in artificial intelligence work on natural language understanding. We will not address them here.

Highlights

In the previous sections we used the tables of observed word usage to explore various models of how systems could name, and thereby give access to, their contents. We devised six general schemes for assigning names to objects. To test the schemes, we approximated user behavior by sampling appropriately from the tables.

Thus, for example, we simulated that word a designer might assign to an object (pick from the table), what word a user might input (pick again from the table), and looked at how frequently the two words were the same. The results were taken to indicate how well a set of spontaneously generated single names for objects can be expected to work for untrained users. This particular example was called the "armchair model." The table is, after all, only a compendium of many people's spontaneous attempts to give the best possible names for these objects. Thus sampling from the table mimics designers picking names from their armchairs.

We used this general technique to explore variants on our basic model of interaction, in which we assume the designer builds in certain connections between names and the system's objects or services. These names were—in some cases—purely random, in other cases, a simulation of a designer's best armchair guess, and in still other cases, chosen more systematically, with a goal of optimality. In all cases, we assumed that the user must "guess" the right name, because we are concerned with untutored users who do not know the system words. The user's choice of words was always mimicked by sampling from the tables.

Of the models and variations, summarized in Appendixes I and II, perhaps the most important are: (a) One Name per Object: Weighted Random; (b) One Name per Object: Optimized; and (c) M Referents for Every Word: Optimized. The first of these was the so called "armchair" naming method just described. Expected results from the armchair method are shown in Figure 4. Note that despite the fact that the four domains differ dramatically in content, data collection style, and subject population, the results are quite consistent. They are all low. In English, two people—the designer and the user—will come up with the same name only about 15% of the time.

These results indicate that the current practice is deficient; it guarantees that it will be difficult for people to tell machines what they want, unless they already know what to say. Note that designers often have other, legitimate motives in assigning names—e.g., lack of ambiguity, memorability, or cuteness. But such naming practices only make the current problem worse; they lead to even less likely names.

The source of the problem is this. There are many names possible for any object, many ways to say the same thing about it, and many different

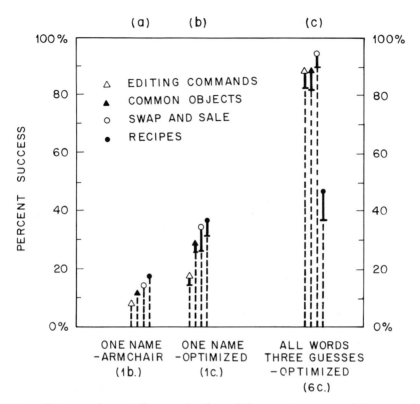

FIGURE 4. **Summary of expected success for three of the most important models presented. Panel (a) shows results for "armchair" model (1b); panel (b) for the best possible single name model (1c); and panel (c) for the model where the system recognizes all words and returns three guesses (6c).**

things to say. Any one person thinks of only one (or a few) of the possibilities. Thus, designers are likely to think of names that few other people think of, not because they are perverse or stupid, but because everyone thinks of names that few others think of. Moreover, since any one person tends to think of only one or a few alternatives, it is not surprising that people greatly overrate the obviousness and adequacy of their own choices.

To improve performance of the "armchair" method we investigated whether experts in one of the content areas could pick better words. We had expert chefs choose keywords for cooking recipes and found essentially no improvement: experts do not seem to do noticeably better picking terms from their armchairs than does anyone else (see discussion of Model 1).

The next model of major interest (Model 1, One Name per Object:

Optimized) substituted objective data for armchair suggestions. It used our data tables to identify the name that was most commonly used for each object. Thus for example, for the operation we referred to as DELETE, the most commonly used term was "omit," with a frequency of 110, and that was the name used in this model. This empirical approach, occasionally advocated by human factors people, achieves the levels of expected success illustrated in Figure 4b.

As discussed in the section on the Optimized version for Model 1 for statistical reasons we can only estimate certain upper and lower bounds on performance here. The lower one indicates how well one could do with a limited amount of empirical data from which to try to pick the best names. The upper one liberally estimates how well one could do with an infinite amount of data. The improvement over the armchair method is substantial, typically almost doubling the hit rate. This makes the point that the best name is a good bit better than the typical name. But is is also clear that even the best is not very good.

It is critical to note that these numbers represent the best one could possibly do by assigning a single name to each object. The problem is not solved by finding the right name. Different people, contexts, and motives give rise to so varied a list of names that no single name, no matter how well chosen can do very well, Figure 5 is a plot of how many names are needed to account for a given percentage of the user attempts, here for the Common Object data. (Here again we have the statistical estimation problem, so the boundary is double.) Note that even 15 words (word types or "aliases") per object account for only 60–80% of the words people apply. That is, even with 15 names stored per object, the computer will miss 20–40% of the time. The lesson clearly is that systems must recognize many, many names.

After considering several other models, we turned the problem around a bit. Instead of starting with system objects and looking for names, the new strategy was to begin with user's words and look for interpretations. That is, try to recognize every possible word that the users generate, and use empirical data to determine what they mean by that word.

A cautionary note: When word meanings are to be surmised from word usage data, there is always risk of the same word having been used in reference to several objects in the system. In such a case, the system would not have a unique guess. (This was actually an implicit problem discussed for several of the earlier models.) In our situations, we limited the system to returning just its single best (Model 5,) or best three (Model 6,) guesses as to what was wanted, given the word the user said. For discussion here we will mention only this last model (Model 6, M Referents for Every Word: Optimized). The results are shown in Figure 4c.

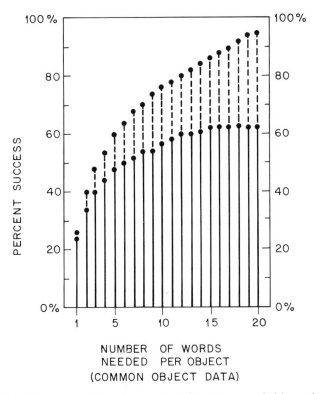

FIGURE 5. **A plot of how many different names (or "aliases") are needed for each object to account for a given percentage of success. This plot is based on data from the Common Object domain. Note that even with 15 words per object, success is only 60–80%. (The exact values are not known because of statistical estimation problems, so ranges are indicated by the dashed lines.)**

This shows what can be achieved if all words are saved by the system, and it uses behavioral data to make the three best guesses as to what was meant. (Here, too, we can only present estimates of upper and lower bounds on performance.)

These results are encouraging. The big improvement is due primarily to the fact that the earlier methods, like most ones in current use, failed to recognize all the rare words. The unfortunate truth is that the majority of words used are "rare" words and so it is a mistake not to reckon with them. Indeed, it turns out (Model 3, section on "Analysis of Precision vs. Popularity of Terms") that less common words tend to be more precise, more discriminating, and so are particularly valuable to have the computer know about.

While we studied quite a wide variety of intermediate models, this last one really is the best. At the risk of belaboring the point, here it is summarized again. It requires: (a) the computer to keep all words that people use; (b) that data be collected on what users use each to refer to; (c) that when some one uses a word, the computer goes to the the data table and looks up the top few candidates as to what its interpretation should be (what the word is most probably referring to); and (d) the computer presents the candidates to the user as a choice, since the interpretations may be in error. This model amounts, really, to a rich, empirically (i.e., behaviorally) defined cross index—without it, performance is near abysmal—with it, potentially quite good.

Further improvements are shown to require the input of multiple words. This would entail means for representing and evaluating relations as well as reference; matters of logic, syntax expression, and language comprehension. Detailed analyses of the complicated usability issues that would be involved is beyond the scope of this paper, but a number of related issues were addresses in the section on "Other Possible Models and Some Limits."

DISCUSSION

We have seen that the object referent of a word cannot be predicted with great accuracy from knowledge of its past referential use. The use of statistical data on reference behavior can improve the choice of a single name by roughly a factor of two over the common procedure in which only a single designer-chosen word is stored as an entry point for each object. Even the best system input–output function of the general kind we have described, while adding another factor of two or so to performance, can be expected to perform at well below perfect reliability. In this section, we touch on some of the reasons for this limitation and suggest some avenues for further exploration of better means for understanding object reference.

Obviously, one of the main difficulties in predicting an intended object from a provided word is synonymy. There are many different words that can refer to the same object. Even though the receiver may know several of them, the communicant (or user) may choose another. It is the long-tailed distribution of the word usage for objects, as illustrated in Figure 1, that is the villain. Unfortunately, however, this is only a part of the problem. Indeed, our upper-bound estimates assume that this problem does not exist, that we know every word that will be used and with what probability each will be applied to every object. Still our best upper-bound predictions are quite error prone. Another part of the problem is polysemy; each word means many different things and can refer to many different objects. In our observed data, words that were frequently used tended to be applied to several objects. Clearly, if a word is used for two or more objects, there is no way that we can guess from the use of that word which unique object is its referent. The more similar the objects in the domain, the more likely it is that a single word will include more than one in its scope (Model 5, section on "Analysis of Weighted Random Version"). In general, then, one might expect that the more similar the objects in a set seem to people, the more difficulty they will have in describing them uniquely and the more difficulty a receiver of their descriptions would have knowing to what they refer.

How might one interpret descriptions of objects more accurately? Observed that in the Common Object experiment, human subjects were able to make a single guess as to the intended object with ober 80% success. This is well above any of our performance estimates for systems based on the statistical information in single-word input–output tables. How do people do this? What other sources of information, either in the input description or in the receiver's mind, are brought to bear? What is needed to build a system that would approach human capacities?

One possibility is that the limits we observed may be due to utilization of only a single word or phrase from the input. Perhaps if multiple words, or the combined meaning implied by conjunctions of several words and their syntactic order, were utilized, much better reference could be achieved. The conjunctive use of words primarily serves to more narrowly specify the object of discourse. This presumably increases the precision of reference and would reduce the chance that we would guess an incorrect object from a provided description. But, under models that assume the comprehender can return several guesses as to the likely object, the recall provided by a full description would not necessarily be much greater than that provided by the unrestricted meaning of its most important word. However, if users can provide many independent words, each an essentially new attempt, the recall rate could be raised substantially, as Model 2 suggests. Thus allowing longer, more complex specifications, and learning how to understand them is one promising direction to be investigated.

Recall also that using experts to generate armchair key words did not work well, at least in the Recipe data where we tested it. We also note that in informal demonstrations, programmer subjects do not fare well in providing a name for a program to be matched by other programmers. Similarly, there have been a number of studies of indexer reliability in bibliographic indexing (Hooper, 1965; Jacoby, 1962; Rodgers, 1961; Tinker, 1966, 1968). These come from quite favorable circumstances, in which the index terms are chosen from restricted vocabularies and the indexers are highly trained. The chances of one indexer choosing the same categories as another, even under these circumstances, are usually disappointingly low. Thus, the chance that professional indexers will agree with the first word entered by an untrained user does not seem to offer a promising route.

A possibility for improvement that seems worthy of further investigation is the study of the structure of the conceptions or mental representations of the objects in a domain to be referenced. The polysemy aspect of the problem arises because two or more objects are not linguistically separable. If we could learn how to group such objects into "super-objects," then we could potentially improve at least our ability to predict which of these super-objects is being sought. Similarly, informal impressions from the swap and sale superordinate data suggest that certain levels of superordination give rise to more consistent naming than others. If means can be found to reveal and represent strong hierarchical structure in the concepts being named, then possibly one can choose the levels or nodes in such structure that are best represented as the objects to be found in a data set. These "super-objects" also might be more amenable to automatic reference by the means we have modeled.

We have generated statistics relevant to this hypothesis, in the double treatment given to the editing terms. The five objects in the Ed5 data are in fact "super-objects," made by condensing the 25 editing operation, text unit pairs into categories involving just the editing operation. A look back at the numbers involved shows a uniform superiority in the retrieval and discrimination of super-objects, over that of the individual constituent objects. In part this is just because there are fewer objects to be dealt with, but our hypothesis is that performance is further enhanced by the lower overall similarity at the super-object level. Suppose a system were built to capitalize on this kind of situation, that is, by first empirically discovering the lowest level of subdivision of a given domain at which natural descriptive language is adequate. Freely chosen keyword entries might well be effective for specifying objects at this level. But the whole problem of retrieval would not yet be solved. Users who wanted to access particular subordinates of these suberobjects would have to be provided with some further mechanism. A hybrid approach in which early stages of specification are done by freely chosen keys and later stages by menu selection seems a promising approach.

An especially important matter that we have so far neglected is the prior probabilities of a user intending various objects. Our data have been aimed at estimation of the input–output pointer functions of words to objects, and were collected with equal numbers of occasions for nomination of key words for each object. In real life, and in a real system, people seek different objects with different frequencies, perhaps with steep distribution functions resembling Zipf's law (Zipf, 1949). Our ability to predict what object a person has in mind by a word could be greatly improved by taking into account its prior, unconditional likelihood of being sought. Again, our informal impressions from the kinds of descriptions provided in the common object data is that human describers and recipients take great advantage of such frequency information. For example, in specifying the Empire State Building as a tall building in midtown Manhattan, the describer probably assumes that the receiver would choose, from the large set of possible objects, the one which is most likely to be the object of specification.

It will also require further work to see how such information could be incorporated in an automated access system. A start would be to consider an adaptive system that keeps track of the frequency with which objects are sought (as well as the frequency with which the particular input is satisfied by a particular output). Then object prior-probability data could be combined with input–output conditional probability data, like that we have investigated, by the use of Bayes's rule (McGee, 1971). This would almost certainly yield much better predictions than our models estimate, or that are currently achievable in available systems. Such a scheme might also take advantage of individual differences in cases where the same people will use the system repeatedly.

There are still other plausible means for circumventing the limitations our models suggest. There are other kinds of data access devices, such as menu-driven systems, query languages, or natural language understanding systems. These approaches all have something to offer, and probably can overcome some of the deficiencies of the pure keyword entry method. But each also has problems of its own. For example, menus cannot list a very large number of alternatives at once, so the desirable feature of turning the recall or production problem into comprehension and choice is limited. (Note also that in menus the user must understand the system's terse descriptions of objects; much as the system has to understand those of the user of key words.) When there are many objects, a menu system must use a sucessive search method that relies in some kind of hierarchical tree or other representation of the relations among the objects. How to do this in a way that leads to correct user choices at each level, and to good overall performance, and to acceptable convenience, are unsolved issues. Query languages generally require users to input well-formed relational algebra or Boolean expressions that are unlike anything seen in the data specifications provided in our password study. Such expressions require the kind of logical thought that is well-known to be extremely difficult for ordinary people (Wason & Johnson-Laird, 1972). Natural-language understanding systems, as so far implemented at least, have yet to deal adequately with the lexical reference problem. They have usually finessed the issue by restricting themselves to very limited domains and limited lexical input. for which they can store a reasonably adequate "hand-tailored" synonym list. We suspect that when such systems are developed for use in real data applications they will have to solve the synonymy and polysemy problems in some of the same ways, e.g., by the use of statistical input—output tables and prior object probabilities that we have been suggesting here. It is probably a mistake to take a too naively optimistic a view of the value of "natural language" input to a computer device. For example, although our human subjects were much better than our model automatic systems at predicting the referent of a description, it is not obvious that their success was based on being able to "understand" the natural language of the input, at least if this is taken to mean successful syntactic parsing, etc. We believe that it is also possible that human success is based largely on statistical knowledge of the likelihood of objects and the likely referents of words, and that this is a matter that could be incorporated into a system without it doing highly intelligent natural language understanding.

SUMMARY

The data we have collected on people describing objects have allowed us to estimate the likely performance of several mechanisms for understanding the references of words to information objects. We have found that input—output functions based on normative naming behavior of users will work much better than systems based on a single name provided designers. We have also shown that a system that made several best guesses, and/or allowed the user to make several tries, and then returned a menulike set of guesses to be chosen among, would substantially improve performance beyond current popular methods. The best approach was to focus on what the user brings to the interaction, namely a great variety of words, and for each word have the system make one or more best guesses as to what the user meant.

But such a system for understanding references will still not perform nearly as well as a human receiver would. Thus, this model of the process of reference certainly does not fully capture what goes on in people's minds. Thus, we are clearly not yet ready to hazard a theory of how humans succeed as well as they do in this task, or to propose an automated method that would do as well or better. However, we believe that the evidence and analyses reviewed here lead to some promising suggestions for further exploration in both regards.

APPENDIX I
THE FORMAL STRUCTURES FOR THE SIX MODELS

The constraints critical to the definition of each model are marked with asterisks. (Note that these constraints are to be met whenever possible, and for simplicity, the numbers below assume that it is always possible.)

Model	Description	Words Used		Objects Referenced		Total Word–Object Pairs
		Per Object	Total	Per Word	Total	
1	One Name Per Object	1^a	$\leq R$	$\leq C$	C^a	C
2	Several Names Per Object	M^a	$\leq R$	$\leq C$	C^a	M × C
3	Distinct Name for Each Object	1^a	C^a	1^a or 0	C^a	C
4	Distinct Names, Augmented with M Extra Referents	$1 \leq n \leq M$	C^a	M^a or 0	C^a	$M^b \times C$
5	Recognize One Referent for Every Word	$\leq C$	R^a	1^a	C	R
6	M Referents for Every Word	$\leq C$	R^a	M^a	C	M × R

[a] A defining constraint for the model; C = number of columns, R = number of rows

APPENDIX II
SUMMARY OF RECALL PROBABILITIES

Model	Description	Version	M	Ed5 (n = 5)	Ed25 (n = 25)	CmOb (n = 50)	Swap (n = 64)	Recp (n = 188)
						Recall Probabilities		
1	One Name	WGT	1	.074	.106	.117	.142	.182[b]
	Per Object	OPT	1	.151	.194	.257	.258	.312[a]
				.271	.322	.329	.353	.419[b]
				.163	.221	.279	.337	.362[c]
2	Several	WGT	1	.074	.106	.117	.142	.182[b]
	Names Per		2	.144	.205	.210	.253	.332[b]
	Object		3	.211	.295	.284	.337	.445[b]
		OPT	1	.151	.194	.257	.258	.312[a]
				.163	.221	.279	.337	.362[c]
			2	.263	.319	.358	.358	.490[a]
				.281	.369	.406	.496	.564[c]
			3	.365	.424	.420	.452	.578[a]
				.381	.492	.478	.595	.670[c]
3	Distinct	WGT	1	.073	.076	.105	.124	.086[a]
	Name for	OPT	1	.140	.109	.226	.194	.105[a]
	Each Object							
4	Distinct	WGT	1	.067	.076	.109	.127	.084[a]
	Names,		2	.125	.145	.151	.172	.149[a]
	Augmented		3	.175	.212	.181	.196	.200[a]
	with M	OPT	1	.140	.109	.221	.208	.105[a]
	Extra		2	.214	.204	.277	.266	.170[a]
	Referents		3	.266	.287	.306	.296	.220[a]
5	Recognize	WGT	1	.413	.153	.516	.617	.128[b]
	One	OPT	1	.489	.176	.434	.351	.129[a]
	Referent for			.620	.353	.647	.715	.276[b]
	Every Word			.541	.258	.686	.813	.247[c]
6	M Referents	WGT	1	.413	.153	.516	.617	.128[b]
	for Every		2	.656	.264	.646	.743	.207[b]
	Word		3	.811	.358	.713	.795	.266[b]
		OPT	1	.489	.176	.434	.351	.129[a]
				.541	.258	.686	.813	.247[c]
			2	.699	.302	.532	.427	.203[a]
				.761	.416	.819	.919	.373[c]
			3	.830	.405	.577	.465	.259[a]
				.884	.531	.881	.956	.455[c]

[a]Split halves.
[b]One of the several repeat-rate related statistics.
[c]Self-prediction.

APPENDIX III

The success of any pure random model considered in this chapter is given simply by the ratio of t, the number of cells included in the system mapping, to RC, the total number of cells in the matrix. To see this simply, note that any structural constraints we might impose are only defined up to a permutation of the rows and columns. Thus consider any table and its group, i.e., all its variants obtained by row and column permutations. The table that is the cellwise total of all these tables must, for reasons of symmetry, be everywhere the same, and its grand total must be t times the number of tables in the group. The success of any individual table is given by the dot product of the user and system matrices (treating the matrices as vectors, i.e., sum the products of corresponding cells). The total success for the group is the sum of these dot products for the members of the group. But the sum of the dot product of one vector with several others is the dot product of that vector with the sum of the others, so the total success for the group is just the dot product of the user matrix and the total matrix. But since the total matrix is uniform, and dividing by the size of the group, we conclude that the average matrix then is just t/rc times the sum of user matrix. If the user matrix is in relative frequencies, the success is a probability.

REFERENCES

Carroll, J. M. Learning, using, and designing command paradigms. *Human Learning: Journal of Practical Research and Application*, 1982, *1*, 31–62.

Child, J., Bertholle, L., & Beck, S. *Mastering the art of French cooking*, Vol I.. New York: Knopf, 1979.

Claiborne, C. *The New York Times cookbook*. New York: Harper & Row, 1961.

Dumais, S. T., & Laundauer, T. K. Unpublished work, 1981.

Furnas, G. W. Unpublished work, 1981.

Gomez, L. M., & Kraut, R. Unpublished work, 1981.

Herdan, G. *Type token mathematics: A textbook of mathematical linguistics.* Mouton: S-Gravenhage, 1960.

Hooper, R. S. Indexer consistency tests—Origin, measurements, results, and utilization. IBM Washington Systems Center, Bethesda, MD. Paper presented at the 1965 Congress International Federation for Documentation, October 10–15, 1965.

Jacoby, J. *Methodology for indexer reliability tests.* RADC-IN-62-1, Documentation, Inc., Bethesda, MD, March 1962.

Keppel G., & Strand, B. Z. Free association responses to the primary responses and other responses selected from the Palermo-Jenkins norms. In L. Postman and G. Keppel (Eds.), *Norms of word association.* New York: Academic Press, 1970. Pp. 177–187.

Landauer, T. K., Galotti, K. M., & Hartwell, S. H. Natural command names and initial learning: A study of text editing terms. *Communications of the Association of Computing Machinery*, 1983, *26*(7).

Lesk, M. E. Another view. *Datamation*, 1981, *27*(12), 146.

McGee, V. E. *Principles of statistics: Traditional and Bayesian.* New York: Appleton, 1971.

Norman, D. A. The trouble with UNIX. *Datamation*, 1981, *27*(12), 139–150.

Our favorite recipes: Inverness Garden Club. Private printing, 1977–78.

Rodgers, D. J. *A study of intra-indexer consistency.* General Electric Company, Washington, DC, September 1961.

Tinker, J. F. Imprecision in meaning measured by inconsistency of indexing. *American Documentation,* 1966, *17,* 96–102.

Tinker, J. F. Imprecision in indexing (Part II). *American Documentation,* 1968, *19,* 322–330.

Wason, P. C., & Johnson-Laird, P. D. *Psychology of structure and reasoning: Structure and content.* Cambridge, MA: Harvard University Press, 1972.

Zipf, G. K. *Human behavior and the principle of least effort: An introduction to human ecology.* Reading, MA: Addison-Wesley, 1949.

9

Facilitating Multiple-Cue Judgments with Integral Information Displays

TIMOTHY E. GOLDSMITH
ROGER W. SCHVANEVELDT

Computer-based decision support systems are being designed to enhance the human decision-maker's inherent information-processing abilities. Previous research has shown that integrality and separability of stimulus dimensions affect performance in perceptual and cognitive processing of multidimensional stimuli. We found that, in a multiple-cue probability learning (MCPL) task, integral displays (enclosed geometric figures) resulted in significantly better performance than a separable display (bar graph) with both additive and configural cue-criterion relations. A new procedure, based on analysis-of-variance, was developed to produce cue-criterion relations with constant linearity and varying levels of configurality. The findings are potentially applicable to the design of display configurations in computer-based decision systems. At a more theoretical level, this study is a step toward integrating information about general cognitive functions with findings in decision-making.

Considerable attention has been devoted to developing automated information systems that aid human operators in processing large amounts of information. The benefits of these systems have stemmed in large part from their ability to acquire, store, and evaluate vast amounts of information. Computer-based decision systems have traditionally exploited the information handling capabilities of the computer by assigning to the computer particular tasks in the processing of information from data input to the selection of an action. This type of aid has reduced the decision-maker's task load and allowed him to devote more resources to those functions that he performs best. However, the successful allocation of tasks to person and machine in realistic decision environments depends on a knowledge of specific task components and their interactions. Such knowledge is often either unavailable or difficult to obtain for complex situations. In addition, people are sometimes reluctant to relinquish part of their decision-making responsibilities to the computer.

Recently, attention has begun to focus on the computer's role as a means for enhancing the information-processing characteristics of the human decision maker rather than as a tool for lessening the decision maker's task load (Freedy & Johnson, 1982). The computer, in this role, functions to reduce the decision-maker's processing limitations and to extend inherent processing capabilities. The human factors specialist, then, is in a position to offer design principles that are geared toward making computer-based decision systems compatible with the information-processing needs of the human operator. By developing support systems that consider the operator's attentional, perceptual, and knowledge representation systems, the decision-maker is maintained as an integral component of the decision process. Hence, the computer is used to reduce the decision maker's cognitive load rather than his task load. In addition, computer assistance of the general cognitive functions of the operator generally results in support systems that are more domain independent. Thus, less knowledge of specific task characteristics may be required.

Cognitive psychologists have developed a large body of data on the characteristics of the human information-processing system. Much of this information can be used to provide guidelines for designing computer-based decision systems that facilitate the processing of information by the human operator. In this chapter, we discuss some research on the perceptual processing of multidimensional stimuli and show how certain findings may be relevant to the design of computerized display configurations. Specifically, we investigate whether decision performance can be improved by displaying decision-relevant information in a structural form that capitalizes on the inherent characteristics of the human perceptual system. If performance can be enhanced, then one supporting role of the computer could be to arrange information output in a manner that optimizes the human's use of that information. The particular type of decision process that we are concerned with is multiple-cue judgment.

MULTIPLE-CUE JUDGMENT

One type of complex decision problem that frequently occurs in everyday situations is the integration of several pieces of information into a final decision. Typically, the information comes from a variety of different sources. Examples of this task include medical diagnosis, assessment of the financial status of business organizations, prediction of successful oil drilling sites, and evaluation of job applicants. In each case, the decision-maker is confronted with multiple information cues that must be combined into an overall judgment of a situation.

A considerable amount of research in psychology exists on people's ability to integrate and act upon multiple sources of information. An excellent review of much of this work in decision making is provided by Slovic and Lichtenstein (1971). A central focus in multiple-cue judgment has been the modeling of a judge's strategy for weighting and combining information cues. Studies in this area find that people's ability to process multiple information cues is limited and, therefore cue-combination strategies are usually quite simple. People typically combine information sources in a linear manner. As a result, statistical models, such as linear regression, often provide accurate descriptions of how a judge predicts a criterion variable. Simple linear models have been shown to predict experts' judgments across a variety of realistic decision situations (Dawes, 1971; Goldberg, 1968; Hoffman, Slovic, & Rorer, 1968). Despite the accuracy of simple linear models to describe experts' decision strategies, most complex decision situations are assumed to contain non-linearly related information sources. Hence, it is likely that experts' performance can be improved in realistic multiple-cue judgments.

In addition to investigating already acquired judgment strategies, studies have also investigated how well people learn to make multiple-cue judgments. A frequently used paradigm for investigating judgment learning is the multiple-cue probability learning (MCPL) task. Under the MCPL task, a subject receives a combination of information cues, estimates a criterion variable associated with those cues, and then receives the true criterion value as feedback. Information cues are experimentally related to the criterion through a functional relation. Therefore, by learning the function relating cues to criterion, subjects are able to improve their estimates of the criterion values. The result of studies employing MCPL tasks is that people can learn to predict criterion values that are related to information cues in a nonlinear fashion, but they do so much slower and less effectively than linear or additive relations (Brehmer, 1969; Summers & Hammond, 1966).

One way to evaluate a person's performance in a MCPL task is to view the judgment situation through the lens-model framework (Hursch, Hammond, & Hursch, 1964). The lens model provides a technique for describing both the characteristics of the task environment and a judge's use of information in that environment. Regression and correlation statistics are used to evaluate a judge's weighting and combining of cues, and then these measures are compared to the optimum strategy determined by the task environment. For example, a measure of the degree to which a judge views a particular cue as relevant to the judgment criterion is obtained by correlating that cue's values with the subject's estimates of the criterion values.

The MCPL task is assumed to have relevance to a number of realistic decision situations that require people to integrate several sources of information into an inference about an uncertain event. In such tasks, there usually exists a complex functional relation between the multiple information cues and the event being predicted. Decision makers attempt to increase their understanding of this relation and improve their judgments by successively estimating an event and observing its outcome. For example, a stockbroker combines various sources of financial information to evaluate the future of a stock. After receiving feedback from previous judgments, the stockbroker may be better able to recognize a certain pattern of data as diagnostic of a particular outcome.

Often the type of information needed to make judgments is numerical. For instance, a stockbroker in assessing a particular company might consider quarterly earnings, outstanding shares, and sales volume. Frequently, this type of information is presented to a judge in a variety of formats. It is probable, however, that the manner in which a judge perceives multiple information sources influences her ability to accurately integrate the information into an overall decision. Support for this idea comes from research on human perceptual processing.

STIMULUS STRUCTURE

A large amount of research in psychology exists on the perceptual processing of multidimensional stimuli. Historically, the interest in this area has focused on the characteristics of the processing system per se. Recently, however, attention has been directed to the inherent structural properties of stimuli and their effect on perceptual processes. Garner (1970) stimulated interest in this line of research by arguing that more understanding is needed of the nature of stimuli before we can rightly understand the characteristics of the human perceptual system.

A central issue in the research on stimulus structure has been the nature of the relations between dimensions which compose multidimensional stimuli. In reviewing this work, Garner (1974) discussed two major ways in which stimulus dimensions can be related. Stimuli composed of integral dimensions (e.g., hue and saturation of a color) produce a Euclidean metric in direct distance scaling, facilitate the discrimination of stimuli on one dimension when another dimension varies in a correlated manner, and inhibit the discrimination of stimuli on one dimension when another dimension varies in an orthogonal manner. Separable dimensions (e.g., separate vertical bars) produce a city-block metric in direct distance scaling and produce neither facilitation with redundant dimensions nor interference with orthogonal dimensions in a discrimination task. Phenomenologically, integral dimensions appear as an integrated whole, whereas separable dimensions are seen as distinct and separate. In addition, dimensions tend to be integral if the existence of one dimension depends on the existence of another dimension. For example, any one of the dimensions of color (brightness, hue, and saturation) can not exist without values on the other two dimensions.

ELEMENTARY PERCEPTUAL PROCESSES

The study of integral and separable dimensions has traditionally been limited to relatively simple stimuli and elementary perceptual processes. Stimuli are generally composed of physical attributes such as color or geometric form and are often, but not always, limited to two binary dimensions. In discrimination and identification studies, the task is to evaluate stimuli on the basis of one dimension while the stimuli vary along either one or two dimensions. The rule relating stimuli to responses is based on the physical attributes of one of the stimulus dimensions. Since these tasks are generally easy to perform, speeded responses are obtained, and reaction time serves as the primary dependent variable. Classification and similarity scaling tasks do not specify what aspects of the stimuli the subject should use in forming classes and assigning ratings. Instead, the interest is on identifying stimulus structure by the way subjects globally view the stimuli. With integral dimensions, classification is based on the overall similarity structure of the stimuli; with separable dimensions, subjects group stimuli on the basis of a single dimension.

CONCEPT FORMATION AND CLASSIFICATION

Related to the work on perceptual classification are studies of how people form concepts. This work has attempted to understand how people learn and use concept rules (Bourne, 1966; Bruner, Goodnow, & Austin, 1956). The experimental paradigm in these studies represents concept rules in terms of the physical attributes of perceptual objects. Thus, learning a concept is equivalent to learning to identify those physical attributes of the stimuli that belong to the concept. Concepts have usually been defined by simple logical rules (e.g., affirmation, conjunction, conditional) involving one or two dimensions of the stimuli. The rules are often conceptually quite simple, and the stimuli employed are usually composed of a small number of qualitative or categorical dimensions, with typically only two or three values on each dimension. The task is made nontrivial by including stimulus dimensions that are irrelevant to the concept. Studies have investigated people's ability to learn both the relevant dimensions of the stimulus and a rule relating the dimensions to the concept (Haygood & Bourne, 1965).

The principal process involved in concept formation is rule induction. A specific rule defines a relation between a set of information cues and a concept and subjects actively seek this underlying rule through the process of inductive reasoning in order to improve performance. The MCPL task also requires rule induction, but there are some important differences between the two tasks. MCPL tasks use several quantitative information cues, and responses are single-valued numerical estimates of a criterion. In contrast, concept formation studies employ stimulus dimensions and responses that are few in number and qualitative or categorical in nature. Furthermore, in MCPL tasks, stimuli are mapped onto responses in such a way that subjects are unable to completely "know" the rule and achieve perfect performance. This can occur because of complexity in the functional relation or because randomness is introduced into the relation. Subjects in concept formation studies, on the other hand, often reach a point of complete understanding of the rule which is then demonstrated by future perfect performance.

The definition of concepts in terms of dimensionally generated stimuli suggests that the manner in which dimensions interact could affect how easily people learn concept rules. Although this issue has not been systematically investigated, Garner (1976) has discussed several possibilities concerning the effect of stimulus structure on concept behavior. Garner suggested that one way to optimize concept learning might be to separate classes of relevant and irrelevant dimensions. This separation should aid the learner in identifying those dimensions of the stimuli that are relevant to the rule. Given that the relevant dimensions are isolated from the irrelevant dimensions, separable relevant dimensions may be preferable to integral relevant dimensions because concept rules are defined by individual dimensions, and thus it may be advantageous to selectively attend to these same dimensions. On the other hand, integral dimensions may facilitate learning by allowing the relevant dimensions to be viewed as a unified whole, thus emphasizing the relation between dimensions.

Garner (1976) concluded that, overall, unitary dimensions should facilitate concept learning, especially if there is redundancy in the dimensions or if the stimuli contain more relevant than irrelevant dimensions. An important implication of this finding is that, because attention is not focused on the individual dimensions of integral stimuli, people may in fact be learning sets of stimuli rather than actual concept rules. To quote Garner (1976), "it is entirely possible that concepts are not really learned as logically defined dimensional rules, but are learned as individual stimuli, except when there is a single relevant dimension, and in that case the issue is uninteresting" (p. 114).

Related to work on concept formation are more recent studies of classification processes. This research is concerned with how people categorize natural objects (Medin & Schaffer, 1978; Reed, 1972; Rosch, Simpson, & Miller, 1976; Smith & Medin, 1981). Like concept formation studies, stimulus objects in these studies usually consist of a small number of categorical dimensions (typically less than five) with two or three values on each dimension. However, despite the similarity of the stimuli, several important differences exist between classification and concept-formation studies. One of these differences is the manner that stimuli are mapped onto responses. In classification studies, the relation between a class and its members is not ordered by a well-defined rule as with concept rules. Instead, class composition is artificially determined to reflect the probabilistic nature of natural categories. A common way of achieving this is to view each stimulus pattern as a point in multidimensional space. Class membership is then defined in terms of inter- and intraclass distance in the pattern space. This results in all of the dimensions being relevant, usually equally so, to determining class membership.

A second difference between classification and concept-formation studies is that classification tasks require subjects to categorize patterns into one of several (typically two or three) classes rather than learning the characteristics of a single concept. It follows that, correct classification behavior is determined more by the development of appropriate category structures, whereas accurate formation of concepts requires learning a specific rule. And finally, classification and concept-formation studies differ in their focus on the stages of categorization. Classification studies are aimed more at discovering the subjective strategies people have already acquired to classify objects, whereas concept-formation studies focus more on how people learn experimentally defined class rules.

As with concept formation, the relation between stimulus dimensions probably affects people's ability to classify stimulus objects into their appropriate categories. Some support for this idea comes from explanations of how integral stimuli are processed. Facilitation with redundant integral dimensions has previously been explained by assuming that the interstimulus distance in Euclidean psychological space increases; the farther apart stimuli are in similarity space, the easier it is to distinguish between them. Apparently, stimuli that differ on two or more integral dimensions are viewed as differing on a single new dimension, and the psychological distance between these stimuli along this new dimension is greater than along a single attribute. The increased dissimilarity of integral stimuli along a global dimension has been attributed to dimensional redundancy. However, Lockhead and King (1977) suggested that facilitation with integral dimensions is not due to dimensional redundancy per se, but is simply tied to the separation of stimuli in psychological space. Thus, classification rules are based on the similarity structure of stimuli rather than their attribute structure. Furthermore, Lockhead (1979; Monahan & Lockhead, 1977) argued that integral stimuli are initially encoded by a holistic process that considers all relevant dimensions simultaneously.

If integral dimensions do in fact increase interstimulus distance, then classification of stimuli represented by integral dimensions would be easier to classify than stimuli represented by separable stimuli. Integral stimuli would appear more distinct, and thus there would be less confusion about which class is appropriate for a particular stimulus. However, this would likely depend on the actual processes people use to learn class structure. There is considerable evidence to indicate that people learn category structure by forming class prototypes, where a prototype is the average of the stimulus dimensions across all exemplars of a single class (Posner & Keele, 1968, 1970; Reed, 1972). Since the development of a prototype seems to require combining individual stimulus dimensions, separable dimensions, which allow attention to be focused on individual dimensions, may actually be better for prototype abstraction. On the other hand, if prototypes are formed in a more holistic fashion, stimuli with integral dimensions may facilitate this process.

Although the role of stimulus structure on classification processes has not been directly explored, one area of research that relates to this issue is the search for effective means of communicating multivariate data. Traditionally, statistical data have been represented numerically or in standard forms such as bar graphs. Recently, attention has turned to discovering alternative graphical representations of multivariate data (Chernoff, 1973; Wang, 1978). A primary aim of this work is to identify representations that easily allow people to comprehend complex relationships inherent in a set of data. One property of graphical representations that may be beneficial for communicating data is stimulus integrality. The holistic processing that occurs with integral dimensions may facilitate the comprehension of relationships that are defined across several dimensions of the data.

Jacob, Egeth, and Bevan (1976) tested integral and separable representations for communicating multivariate data using a classification task with complex stimuli. In their study, each data point was composed of nine dimensions with each dimension varying along ten values. Integral displays consisted of schematic faces and polygons, and arrays of numeric data formed a separable display. Subjects were shown five prototypes and asked to rapidly sort stimulus cards into piles corresponding to the prototypes. Prototypes were chosen to be well separated in multidimensional space and exemplars of prototypes were selected so that each exemplar was closest in Euclidean distance to its prototype. Since the task used in Jacob et al. allowed subjects to directly compare each stimulus pattern to a prototype by simple visual inspection, subjects were not required to remember class information. Thus, subjects did not need to store category information to perform the task accurately.

The results of the study showed that the integral displays were better than the separable display, and the face was superior to the polygons. Apparently, well-integrated displays, such as the face, facilitate performance with complex as well as simple stimuli. Jacob et al. argued that the face was a good communication mode, not simply because it integrates information, but because people are familiar with processing faces and may use already developed perceptual processes in evaluating schematic faces.

Taken together, the above research indicates that the structure of stimuli plays an important role, not only in elementary perceptual processes, but also in higher-order cognitive operations including classification and choice. More specifically, stimuli composed of integral dimensions appear to facilitate performance across several types of perceptual and cognitive tasks. However, most of the evidence for an integral advantage comes from studies employing somewhat artificial tasks. The purpose of this study is to examine the role of integral dimensions in a situation more relevant to "real world" problems. In particular, we investigate whether integral dimensions will aid performance in multiple-cue judgment. We chose the MCPL task to study this issue because of its assumed relevance to many realistic situations.

Another advantage of using the MCPL paradigm is that well-defined methods exist for measuring characteristics of the task that appear to be important in studies of stimulus structure. With the lens model framework, for example, it is possible to obtain measures of cue (dimensional) redundancy and relevancy. Redundancy between two information cues is measured by the correlation between the values on the two cues, and cue relevancy is measured by the correlation between the cue values and the true criterion values.

Decision-relevant information is mainly conceptual rather than perceptual in nature, and therefore is typically displayed in numeric or narrative form. However, by presenting numeric information cues graphically, the role of stimulus integrality on judgment can be investigated. As mentioned previously, people are poor at combining multiple sources of information, especially when the information is related to a criterion in a nonlinear manner. It is possible that the integration of information that occurs perceptually with integral dimensions may also facilitate the integration of information at a higher cognitive level. Therefore, we hypothesize that part of a person's processing load will be reduced by perceiving multiple information cues represented with integral dimensions. In the following series of experiments, we investigate the effect of integral and separable information displays on subjects' ability to combine and use multiple sources of information in a MCPL task.

GENERAL PROCEDURE

This section describes those procedures common to all of the following experiments.

Subjects and Design

Information display configurations and cue-criterion relations were factorially related in a between-subjects design. In all experiments, 15 subjects were randomly assigned to each condition. Subjects were introductory psychology students at New Mexico State University who participated in partial fulfillment of a course of requirement.

Cue-Criterion Relations

The manner in which information cues are related to a criterion has been shown to be important in previous MCPL studies. People find it particularly difficult to use cues that are nonlinearly related to a criterion. For this reason we examined display configurations with both linear and nonlinear relations. Previous studies have investigated the complexity of cue-criterion relations by varying either the linear–curvilinear or the additive–configural components of the relation (e.g., Brehmer, 1969; Summers & Hammond, 1966). A curvilinear relation is produced by including exponential, trigonometric, or other nonlinear terms in the function relating cues to criterion. A configural relation exists when the interpretation of one cue varies with the value of other cues. Consequently, configural relations require using the interdependency among the cues to correctly predict a criterion.

The multiple correlation between a set of information cues and a criterion has previously been used as a measure of both the linear-curvilinear and the additive–configural characteristics of the task. A problem with this approach is that linear and configural components cannot be separated on the basis of this measure alone. Instead, we define linearity and configurality by means of an analysis-of-variance model with orthogonal polynomials used to analyze the linear components. An analysis-of-variance, with cues as the only factors in the design, is performed on the set of criterion values. Linearity is defined by the proportion of total variation that occurs in the linear components of the cue variables and their interactions. Configurality is defined by the proportion of total variation that occurs in the interactions between the cue variables. Hence, with the analysis-of-variance technique to measure nonlinearity of a cue-criterion relation, configurality can be varied independently of linearity. Based on earlier findings that showed task difficulty increased with decreasing values of the multiple correlation between cues and criterion, we expect increasing levels of configurality to result in increasing the difficulty of learning the cue-criterion relation.

The analysis-of-variance approach was used to generate various types of functions for relating information cues to a criterion. Because we were mainly interested in the the effect of varying configurality, we held linearity to approximately one for all relations. We maximized linearity by choosing as the original cue values a set of linear coefficients of orthogonal polynomials. Configurality was then varied by combining the cue values with different functions. In each case, the cue values were orthogonally combined resulting in a completely uncorrelated set of cue-value combinations. For example, one set of functions [Eqs. (1), (2), and (3)], used two cues (x and y) with 10 values for each cue. The cue values consisted of the set of 10 linear orthogonal coefficients. Sets of 100 criterion values were produced by combining the cue values with the following functions:

$$\text{Additive:} \quad C = (x + y + 18) / 2 \tag{1}$$

$$\text{Configural:} \quad C = (x \times y + 81) / 9 \tag{2}$$

$$\text{Mixed:} \quad C = (2 \times x \times y + 9 \times x + 9 \times y + 162) / 27 \tag{3}$$

An additive relation [Eq. (1)] was produced by simply adding the values of the two cues, x and y. Figure 1 shows the graph of each combination of cue values plotted against the criterion values for the additive relation. With this relation, each level of one cue is completely additive with respect to the levels of the other cue. Hence, increasing values of cue x always correspond to increasing criterion values for each level of cue y. Since none of the variation occurs in the interaction of the two cues, an analysis-of-variance on the criterion values resulted in a configurality measure of zero for the additive relation.

ADDITIVE RELATION

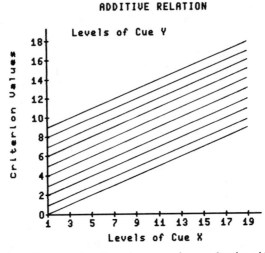

FIGURE 1. **Criterion values as a function of cue** *x* **and cue** *y* **for the additive cue-criterion relation.**

A configural relation [Eq. (2)] resulted from multiplying the values of cues *x* and *y*. Figure 2 shows how the configural relation looks graphically. Here, the two cues are completely interdependent. Increasing values of cue *x* correspond to increasing criterion values for half of the levels of cue *y*, but increasing values of cue *x* correspond to decreasing criterion

CONFIGURAL RELATION

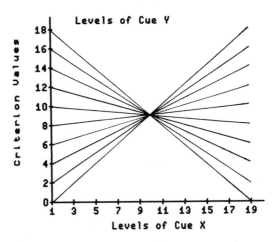

FIGURE 2. **Criterion values as a function of cue** *x* **and cue** *y* **for the configural cue-criterion relation.**

MIXED RELATION

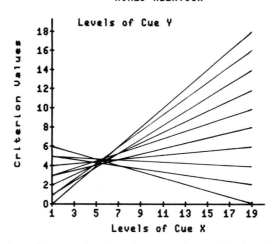

FIGURE 3.   **Criterion values as a function of cue *x* and cue *y* for the mixed cue-criterion relation.**

values for the other values of cue *y*. In other words, all of the variation occurred in the interaction of the two cues. Hence, configurality was equal to one.

A mixed relation [Eq. (3)] was produced by adding together the sum and the product of the cue values. As seen in Figure 3, the cue values were only partially interdependent. Increasing values of cue *x* correspond to increasing criterion values for 7 values of cue *y*, and increasing values of cue *x* correspond to decreasing criterion values for 3 values of cue *y*. Thus, about half of the total variation occurred in the interaction of the two cues and half occurred in the main effects. An analysis-of-variance of the criterion values showed configurality to be .46 for this relation.

The constant terms in each equation were used simply to produce an identical range of criterion values for all three relations, in this case 0 to 18. When necessary, the results of the functions were rounded to the nearest integer. The actual cue values of *x* and *y* displayed to the subjects were produced by adding 10 to the original cue values. This produced the set of values: 1, 3, 5, 7, 9, 11, 13, 15, 17, 19. Since the results of an analysis-of-variance are not affected by adding a constant to the values of the independent variables, the measures of configurality are not changed.

A second set of functions was derived for displaying 3 cues (*x*, *y*, and *z*). In this case, each cue took on five values producing a set of 125 criterion values. The following equations were used to combine the original set of five linear orthogonal coefficients:

$$\text{Additive:} \quad C = x + y + z + 36 \tag{4}$$

$$\text{Configural:} \quad C = (3 \times x \times y \times z + 144) / 4 \tag{5}$$

These functions resulted in configurality values of zero and one for the additive and configural relations, respectively. Criterion values ranged from 30 to 42. The actual cue values displayed to the subjects were obtained by adding three to the original values resulting in the set: 1, 2, 3, 4, and 5.

A final set of functions was derived for 3 cues (x, y, and z), but this time one of the cues (z) was completely irrelevant to the set of criterion values. The same set of cue values was used as in the preceding set of functions. The cue-criterion functions were:

$$\text{Additive:} \quad C = x + y + 34 \tag{6}$$

$$\text{Configural:} \quad C = x \times y + 34 \tag{7}$$

Configurality values produced by these functions were zero for the additive relation and one for the configural relation. Criterion values ranged from 30 to 38. Although cue z was not used in determining the criterion values, all possible combinations of all three cues were displayed to the subjects.

Apparatus

A Terak 8510 microcomputer controlled the experiment and collected the data. The information displays were graphically presented on the Terak CRT.

Information Displays

The integral information displays are described separately for each experiment. The separable display always consisted of either two or three bars in a bar graph, depending on the number of information cues. Variation in each cue value was mapped onto the height of one of the bars. The bars were 4 mm wide and ranged in height from 5 mm to 95 mm. Bar graphs were chosen to represent information in a separable display because they are clearly separable dimensions and also because they are commonly used for graphically displaying data.

Instructions

Instructions were presented to subjects on the Terak CRT. Subjects were shown an example of the type of display they would be using. They were told that each display was associated with a number and that this number was determined by a constant relation between the displays and the values. Subjects were told that their task was to learn this relation in order to improve their predictions of these values. In the case of the bar graph, subjects were informed that the heights of the bars were the cue values that determined the display's criterion value. Subjects were encouraged to respond quickly and to reduce their error scores throughout the experiment.

Procedure

A block of trials consisted of all of the cue value combinations for a particular cue-criterion relation (either 100 or 125 trials). The presentation order of the displays within each block of trials was randomized for each subject.

Subjects were seated at normal viewing distance from the CRT and keyboard. On each trial, a subject saw the information display, entered a response on the keyboard, and then observed the true criterion value appear immediately below his response. All information remained on the screen for 2 sec at which time the screen was cleared and the next display appeared. A warning tone reminded subjects to respond 7 sec (except Experiment 1, which was 11 sec) after the display appeared. Performance feedback consisting of a standardized mean-square error measure was presented for every 10 trials.

Data Analysis

Performance was measured by the correlation between subjects' estimates of the criterion values and the true criterion values. Subjects' estimates were individually correlated with the criterion values for each block of trials and then transformed to Fisher's Z coefficients. The data analysis was performed on these transformed values.

Experiment 1

Experiment 1 presented subjects with two blocks of trials from one of the cue-criterion relations of Eqs. (1) through (3). Ninety subjects participated in a two types of display configurations by three cue-criterion relations factorial design.

Integral Displays

Rectangles were chosen to represent the two information cues, x and y, in an integral display. Variation in cue values was mapped onto the

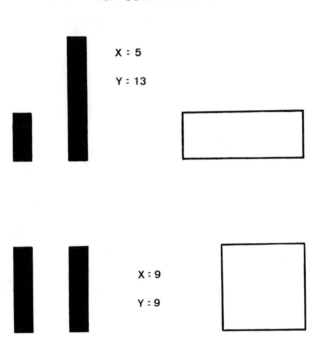

FIGURE 4. **Examples of bar graph and rectangle displays for cues *x* and *y*.**

height and width of the rectangles. The height and width of rectangles
have previously been shown to be integral dimensions (e.g., Felfody,
1974). Subjects were told that the relevant dimensions were the height
and width of the rectangles. The rectangles varied in height and width
from 5 to 95 mm. Figure 4 shows examples of the bar graph and rectangle displays used in Experiment 1.

Results and Discussion

Table 1 presents the mean achievement correlation in Fisher's Z
scores for each condition. The data were analyzed in a 2 Displays × 3
Relations × 2 Blocks of trials analysis-of-variance. The analysis revealed
significant main effects for displays, $F(1,84) = 8.00$, $p < .01$, relations, F
$(2,84) = 176.96$, $p < .001$, and blocks of trials, $F(1,84) = 119.55$,
$p < .001$. Overall, the rectangle display resulted in better performance
than the bar graph display. Thus, integral dimensions appear to aid in
the integration of separate pieces of information in multiple-cue judgment. As expected, performance decreased as the configurality of the

TABLE 1
Mean Achievement Correlations in Fisher's Z Scores for Experiment 1

| | Cue-Criterion Relation | | | | | |
| | Additive | | Mixed | | Configural | |
Blocks	Bar	Rec	Bar	Rec	Bar	Rec
1	.96	1.32	.82	.95	.11	.19
2	1.54	1.53	1.15	1.22	.23	.44
Mean	1.25	1.43	.98	1.08	.17	.32

Note. Bar refers to the bar graph display, and Rec refers to the rectangle display.

cue-criterion relations increased. Also, performance significantly increased between blocks of trials.

Significant interactions also occurred for Blocks of Trials × Relations, $F(2,84) = 5.33, p < .01$, and Blocks of Trials × Relations × Displays, $F(2, 84) = 7.22, p < .01$. The advantage of the rectangle display varied with configurality and with amount of prior practice. The advantage was greatest in Block 1 for the additive relation, $t(84) = 5.49, p < .01$, and Block 2 for the configural relation, $t(84) = 3.20, p < .05$. No other display effects reached standard significance levels. These results suggest that the facilitation provided by an integral display is affected by both the functional relation between the information cues and the criterion as well as the level of practice with the relation.

Subjects had little difficulty learning to combine information in an environment that was purely linear and nonconfigural. Performance was high in the additive relation by the end of Block 1 and increased minimally between blocks. A ceiling effect may have produced the insignificant display effects observed in the second block with the additive relation. In contrast, little learning had occurred with the configural relation even after 100 trials. Not until the second block of trials did performance begin to improve, and it is here that the integral display was significantly better. A floor effect may have resulted in the insignificant display effects in the first block. Taken together these findings suggest that integral displays may be most beneficial during a period of significant learning of the cue-criterion relation. Furthermore, when subjects are either in a state of complete ignorance about the cue-criterion relation or are at an asymptotic level of performance, the benefit of an integral display appears to be minimal. We explore this issue further in Experiment 4.

Facilitation of information-processing tasks with integral dimensions has been explained by an increase in the psychological distance between integral stimuli. Integral dimensions combine to produce a unique stimulus, and it is the global processing of this new stimulus that aids task performance. This explanation seems plausible for tasks that require people to discriminate between stimuli, especially when the appropriate classes are originally determined by overall distance in Euclidean space. However, in the MCPL task stimuli are mapped onto responses by an explicit rule that governs how dimensions (cues) are to be combined. It is possible that, with this task, people need to attend to the same dimensional structure of the displays that are used to define the display and accordingly used in the cue-criterion rule. Even with integral stimuli, it is possible to analyze dimensions separately. Therefore, subjects with the rectangle display may have analyzed the rectangle into its height and width and used this information to form a rule. If so, then the rectangle advantage would imply that it is easier to learn a cue-criterion relation by attending to the height and width of rectangles than to the heights of bars in a bar graph. On the other hand, the benefit of the rectangle display may arise from subjects processing a new global dimension of the display. In this case, the actual attributes composing the integral display may not be important, although the cue-criterion rule uses these values.

One way to test whether subjects are using the individual dimensions of the integral display to form rules, or instead are focusing on a global dimension of the display, is to obscure the relevant display dimensions. If the integral display advantage comes from an analysis of integral dimensions, then masking these dimensions should result in poorer performance. If, on the other hand, subjects are aided primarily by initially processing a new global dimension, then attending to the relevant cue dimensions should not be important. Experiment 2 was designed to investigate this question.

In Experiment 2, we also increase the number of information cues from two to three. Most realistic multiple-cue judgments require people to consider more than two sources of information. It is likely that a judge's cognitive load increases as the number of separate pieces of information that he must consider increases, especially in a task where all of the dimensions are relevant to the criterion. If integral displays assist judgment by reducing cognitive load, then the advantage of an integrated display may actually increase as the number of information cues increases.

Experiment 2

Experiment 2 used the additive and configural relations for three cues given in Eqs. (4) and (5). Two blocks of trials were given to ninety subjects.

Integral Displays

Two new integral displays were tested in this experiment. These displays were produced with a method described by Jacob, Egeth, and Bevan (1976). With this method, variations in cue values are represented by the lengths of equally spaced radii emanating from a common center. A multidimensional polygon is created by connecting the end points of adjacent radii. So, for three cues the resulting figure is a triangle. In one of the displays, we left the radii in the figure, and in another display we removed the radii leaving only a triangle. Both displays varied in width from 25 to 125 mm and in height from 17 to 85 mm. In contrast to the previous experiment, subjects were not informed of the relevant cue dimensions of the integral displays. Instead, the instructions encouraged subjects to focus on the overall shape or form of the figure to learn to predict the criterion values. The separable display consisted of three bars in a bar graph. As before, subjects were told that the heights of the bars were the relevant cues. Examples of the triangle displays and bar graphs used in Experiment 2 are shown in Figure 5.

Results and Discussion

Performance scores for Experiment 2 are presented in Table 2. A 3 Displays × 2 Relations × 2 Blocks of Trials analysis-of-variance showed significant main effects for displays, $F(2,84) = 5.56, p < .01$, relations, $F(1,84) = 349.01, p < .001$, and blocks of trials, $F(1,84) = 56.71, p < .001$. Both types of triangle displays resulted in better performance than the bar graph for both types of cue-criterion relations. The facilitation from the triangle display was greater for the additive than the configural relation, although the Displays × Relations interaction failed to reach significance, $F(2,84) = 2.79, p < .10$. A significant Relations × Blocks of Trials interaction $F(1,84) = 14.36, p < .001$ again reflected the greater increase in performance across blocks of trials for the additive than configural relation.

The integral display facilitated performance even when subjects were not informed of the relevant cue dimensions but instead were instructed to focus on the overall form of the display. This was true even in a display that did not explicitly show the actual cue dimensions. Therefore, it seems likely that the integral display facilitation did not depend on subjects attending to the individual dimensions of the display. It is possible, though, that subjects were aware of changes in physical attributes of the display that were correlated with changes in the actual cues. For instance, the size of the points of the triangles was a salient cue that directly varied with the lengths of the radii. However, discussions with some of the subjects after the experiment indicated that most were not

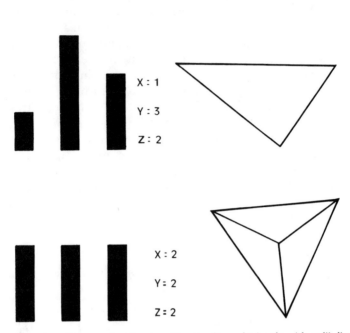

FIGURE 5. **Examples of bar graph, triangle without radii, and triangle with radii displays for cues *x, y,* and *z*.**

TABLE 2
Mean Achievement Correlations in Fisher's *Z* Scores for Experiment 2

	Cue-Criterion Relation					
	Additive			*Configural*		
Blocks	*Bar*	*Tri1*	*Tri2*	*Bar*	*Tri1*	*Tri2*
1	.93	1.09	1.36	.07	.06	.07
2	1.26	1.51	1.70	.10	.22	.24
Mean	1.09	1.30	1.53	.08	.14	.16

Note. Bar refers to the bar graph display, Tri1 refers to the triangle display with radii, and Tri2 refers to the triangle display without radii.

attending to a particular set of attributes. Thus, the facilitation apparently resulted from an increased separation of the integral displays in psychological space. If so, then complex rule-induction processes may actually be more similar to the development of appropriate category structures rather than to the formation of cue combination strategies. This view is supported by evidence that people often judge an object according to how representative it is of a class of objects (Kahneman & Tversky, 1972). For example, a clinician might evaluate whether an MMPI profile indicates psychosis by the overall similarity of the test profile to a prototypical psychotic patient. A rule process, on the other hand, implies that the clinician is intentionally combining the profile's scale values to reach a diagnosis.

Increasing the number of information cues from two to three did not greatly change overall performance. It is possible that, with the cue-criterion relations we used, a greater increase in the number of information sources is required to significantly increase load.

In the previous experiments, each information cue was equally relevant to predicting the criterion. In most realistic decision situations, however, some information sources are more relevant to judging a criterion than others. In this case, the judge must weight the items of information differentially according to their relevancy. Often, however, a judge does not know which sources are more predictive. In fact, people may sometimes unwittingly consider information that is totally unrelated to the event they are attempting to predict. The purpose of the next experiment is to examine integral displays in a MCPL task with both relevant and irrelevant cues.

As mentioned earlier, Garner (1976) suggested that integral dimensions in a concept-formation task are most effective when all of the dimensions of the stimuli are relevant. The same claim would also seem to be true for an MCPL task. With a separable display, selective attention to individual dimensions is more likely, and accordingly, subjects are more likely to identify one of the dimensions as irrelevant. This, in turn, should reduce their task to dealing with only two cues. In contrast, with an integral display, the essence of individual dimensions is attenuated, and the irrelevant cue is less likely to be identified. In this case, an irrelevant cue might harm the integration of information.

Experiment 3

Experiment 3 compared the triangle without inside radii to the bar graph display from Experiment 2. The relevant cue-criterion relations were Eqs. (4) and (5) from Experiment 2. The irrelevant relations were

Eqs. (6) and (7). Subjects assigned to the irrelevant relations were not told that one of the cues was irrelevant. Two blocks of trials were given to ninety subjects.

Results and Discussion

The results from Experiment 3 are shown in Table 3. The data were analyzed in a 2 Displays × 2 Relations × 2 Cue Relevancies × 2 Blocks of Trials analysis-of-variance. Significant main effects were found for displays, $F(1,112) = 11.41, p < .01$; relations, $F(1,112) = 562.40, p < .001$; cue relevancies, $F(1,112) = 22.94, p < .001$; and blocks of trials, $F(1,112) = 100.57, p < .001$. The integral display was superior to the separable display with display differences most pronounced with the relevant cue-criterion relations. However, even with the irrelevant relations, the integral display was somewhat better than the bar graph overall. A significant Relations × Cue Relevancies interaction, $F(1,112) = 38.16, p < .001$, showed that the difference between relevant and irrelevant relations was most pronounced with the additive relation. Apparently adding the irrelevant cue to a fairly easy relation seriously reduced performance. However, with the configural relation, subjects were already performing quite poorly and adding the irrelevant cue did not significantly hinder performance.

Since subjects in the separable condition were attending to the individual dimensions, they eventually could have identified and then ignored the meaningless cue, at least with the additive relation. Some of the subjects using bar graphs reported following this strategy. Subjects using the integral display, on the other hand, were not informed about the true dimensions and so would have been less likely to ignore the irrelevant dimension. This may have accounted for the lack of display

TABLE 3
Mean Achievement Correlations in Fisher's Z Scores for Experiment 3

	Cue-Criterion Relation							
	Additive				Configural			
	Relevant		Irrelevant		Relevant		Irrelevant	
Blocks	Bar	Tri2	Bar	Tri2	Bar	Tri2	Bar	Tri2
1	1.05	1.28	.73	.74	.03	.11	.05	.18
2	1.34	1.59	1.07	.97	.07	.34	.17	.36
Mean	1.19	1.43	.90	.86	.05	.22	.11	.27

Note. Bar refers to the bar graph display, and Tri2 refers to the triangle display without radii. Relevant refers to all relevant cues and Irrelevant refers to one irrelevant cue.

differences with the additive irrelevant relation. With the configural relation, though, the task was sufficiently complex that subjects using the separable display never identified a cue as irrelevant. Furthermore, the integral display was better than the separable display with the configural relation. Apparently, then, integral displays can still facilitate information integration, even with irrelevant cues, as long as the task is reasonably difficult.

The final experiment addresses the role of floor and ceiling effects on the integral display advantage. We hypothesized earlier that facilitation from an integral display is greatest during periods of significant learning and that, when subjects either have no information about the relation or are at a level of asymptotic performance, display effects disappear. The final experiment further investigates this idea by extending the learning period with the configural relation. Remember, with the configural relation after the first block of trials, subjects have little knowledge of the cue-criterion relation. By extending the number of blocks of trials, subjects should eventually improve their understanding of the relation and consequently larger display effects should be found.

Experiment 4

Experiment 4 compared the triangle display with inside radii to the bar graphs under the configural relation of Eq. (4). Two blocks of trials were presented to thirty subjects on three separate days for a total of 750 trials per subject.

Results and Discussion

A 2 Displays × 3 Days × 2 Blocks of Trials analysis of variance showed significant main effects for displays, $F(1,28) = 30.59, p < .001$; days, $F(2,56) = 58.83, p < .001$; and blocks of trials, $F(1,28) = 25.39, p < .001$. The triangle display was significantly superior to the bar graphs for representing configural information. Significant Displays × Days, $F(2,56) = 7.81, p < .01$, and Displays × Blocks of Trials, $F(1,28) = 4.22, p < .05$ interactions resulted from an increasing difference in performance between the integral and separable displays across both blocks of trials and days. Figure 6 shows how the advantage of the triangle display steadily increased from a difference in achievement scores of .07 in Block 1 to .48 by Block 6. Subjects with the integral display learned to predict the criterion values considerably better than those with the separable display by the end of the experiment.

The results from Experiment 4 support our hypothesis that integral displays are most effective during periods of active learning. At the beginning of the task, subjects had little knowledge of the difficult con-

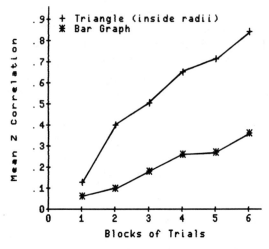

FIGURE 6. **Mean Z correlation across blocks of trials for bar graph and triangle displays.**

figural relation. As performance improved with increasing amounts of training, the advantage of the integral display increased. This advantage continued to increase over the six blocks of trials, and there was no indication that performance had asymptoted. Apparently, integral displays are especially effective when subjects are actively acquiring knowledge about a cue-criterion relation.

CONCLUSIONS

The manner in which stimulus dimensions interact has previously been shown to play an important role in certain types of information-processing tasks. The present study investigated the effect of stimulus structure on a higher-order cognitive process. Our results indicate that the manner in which stimulus dimensions interact can affect complex decision processes such as multiple-cue judgment. The combined findings from our experiments suggest that the integration and use of multiple sources of information can be facilitated by presenting information cues to a judge in a display configuration with integral dimensions.

Applications to Computer-Based Decision Systems

Recently attention has begun to focus on developing computer-based decision systems that meet the information-processing needs of the user (Benbasat & Taylor, 1982). The findings from the present study are potentially applicable to the design of these systems. Based on our findings, one role of the computer in decision support systems could be to arrange information cues into an integrated display configuration. The conditions under which we tested integral displays appear to be important for many realistic decision environments. Fortunately, the advantage of integral over separable displays was quite robust under these conditions. For example, integral displays improved performance with both simple additive and complex configural cue-criterion relations. This is especially important because people are essentially linear information processors, but most realistic decision tasks are assumed to contain configurally related information. Also, integral displays were shown to be at least as good or better than separable displays when one of the information cues was irrelevant. Again, judges in realistic tasks do not always know which sources of information are most relevant to the criterion they are predicting.

An ever increasing number of people receive decision-relevant information via computer terminals, many of which have graphics capabilities. Most often this information is presented in a numeric, piecemeal fashion that requires the user to process it sequentially. Substantial improvements in decision performance may be possible by having the computer arrange multiple sources of information in an integrated output configuration. In this way, a user's general ability to transform information into decisions can be enhanced by capitalizing on the inherent characteristics of the human information-processing system.

Although, the integral display advantage appeared quite robust in the present study, the effect of integrated displays in decision making still needs further investigation. In an earlier experiment not reported here, we found no difference between bar graphs and rectangles with a configural relation. This result was apparently due to the unlikely assignment of some exceptional learners to the bar graph condition. However, more research is still needed to determine the conditions under which integrated displays are optimal.

One issue needing further investigation is whether certain types of integrated displays are better than others. One type of display that has been evaluated for representing complex multidimensional information is "Chernoff" faces (Chernoff, 1973; Jacob, Egeth, & Bevan, 1976; Naveh-Benjamin & Pachella, 1982). This display technique represents variation in each dimension of a set of data by variation in a characteristic of a cartoon face. A potential advantage of the face display is that people are already good at recognizing faces, and so complex multivariate data can be easily communicated by particular facial expressions.

A possible disadvantage of the face display is that certain characteristics of the face (e.g., mouth) appear more salient and consequently will attract more attention than others. This may be troublesome if all dimensions of the data are equally relevant to the task. Recently, Naveh-Benjamin and Pachella (1982) showed that people could not ignore irrelevant information when judging the similarity of two faces if the irrelevant information distinguished between the faces and was represented by salient features. Also, in some pilot testing, we found faces to be inferior to a bar graph display for representing multiple information cues. This contrasts with the findings of Jacob, Egeth, and Bevan (1976) discussed earlier. An important difference between the face display and polygons, which facilitated performance, is that variation in cues is mapped onto differing dimensions (characteristics) of a face, while polygons represent variation with uniform dimensions. Perhaps it is important to have integral displays that preserve the comparative values on different dimensions when information integration is involved.

Another question that needs further research is the effectiveness of integral displays for representing information sources that vary along measurement scales. For example, it is not clear whether integral displays would be beneficial when both quantitative and qualitative information must be combined into a single judgment. It is likely though, that information at least at the ordinal level could be successfully combined in an integral display with information at higher measurement scales.

A distinction in multiple-cue judgment exists between the acquisition of decision strategies and the application of already acquired knowledge. The MCPL task is primarily concerned with the learning phase of judgment. Additional research is needed to examine the effect of integrated displays in situations where a person has already developed strategies for a particular problem domain. As mentioned earlier, performance with expert judges is often less than optimal in realistic decision tasks. Accordingly, judges may continue to alter and refine their strategies as they gain experience. Under these conditions, integrated displays are likely to aid performance.

Theoretical Implications

A recent tendency in decision-making research is to emphasize the importance of general cognitive processes and structures for understanding human decision behavior (Einhorn & Hogarth, 1981; Wallsten, 1980). The present study adds to this line of research by illustrating one way in which perceptual processes interact with higher-order decision processes. Future research in this direction could extend our understanding of human decision-making by focusing on how other cognitive systems influence and interact with decision processes.

There is a considerable amount of evidence indicating that people's decision-making abilities are prone to systematic errors and biases. For example, people revise estimates based on probabilistic data conservatively (Edwards, 1968), fail to take base-rate information into account (Bar-Hillel, 1980), and fail to obey the laws of expected utility theory when considering risky decisions (Kahneman & Tversky, 1979). Attempts to aid decision-makers have usually taken the form of simplifying or eliminating some aspect of the decision process. The results of the present study suggest that judgment errors may be reduced by changing the manner in which decision information is perceived. Perhaps alternative ways of displaying information would assist decision makers in areas other than multipe-cue judgment. For instance, representing base-rate information together with individuating information in an integrated display may make it more likely that people will consider base rates when estimating probabilities.

REFERENCES

Bar-Hillel, M. The base-rate fallacy in probability judgments. *Acta Psychologica*, 1980, *44*, 211–233.

Benbasat, I., & Taylor, R. N. Behavioral aspects of information processing for the design of management information systems. *IEEE Transactions on Systems, Man, and Cybernetics*, 1982, *SMC-12*, 439–450.

Bourne, L. E. Jr. *Human conceptual behavior*. Boston: Allyn & Bacon, 1966.

Brehmer, B. Cognitive dependence on additive and configural cue-criterion relations. *The American Journal of Psychology*, 1969, *82*, 490–503.

Bruner, J. S., Goodnow, J. J., & Austin, G. A. *A study of thinking*. New York: Wiley, 1956.

Chernoff, H. The use of faces to represent points in k-dimensional space graphically. *Journal of the American Statistical Association*, 1973, *68*, 361–368.

Dawes, R. M. A case study of graduate admissions: Application of three principles of human decision making. *American Psychologist*, 1971, *26*, 180–188.

Edwards, W. Conservatism in human information processing. In B. Kleinmuntz (Ed.), *Formal representation of human judgment*. New York: Wiley, 1968.

Einhorn, H. J., & Hogarth, R. M. Behavioral decision theory: Processes of judgment and choice. *Annual Review of Psychology*, 1981, *32*, 53–88.

Felfoldy, G. L. Repetition effects in choice reaction time to multidimensional stimuli. *Perception & Psychophysics*, 1974, *15*, 453–459.

Freedy, A., & Johnson, E. M. Human factor issues in computer management of information for decisionmaking. *IEEE Transactions on Systems, Man, and Cybernetics*, 1982, *SMC-12*, 437–438.

Garner, W. R. The stimulus in information processing. *American Psychologist*, 1970, *25*, 350–358.

Garner, W. R. *The processing of information and structure*. Hillsdale, NJ: Erlbaum, 1974.

Garner, W. R. Interaction of stimulus dimensions in concept and choice processes. *Cognitive Psychology*, 1976, *8*, 98–123.

Goldberg, L. R. Simple models or simple processes? Some research on clinical judgments. *American Psychologist*, 1968, *23*, 483–496.

Haygood, R. C., & Bourne, L. E. Jr. Attribute- and rule-learning aspects of conceptual behavior. *Psychological Review*, 1965, 72, 175–195.

Hoffman, P. J., Slovic, P., & Rorer, L. G. An analysis-of-variance model for the assessment of configural cue utilization in clinical judgment. *Psychological Bulletin*, 1968, 69, 338–349.

Hursch, C. J., Hammond, K. R., & Hursch, J. L. Some methodological considerations in multiple-cue probability studies. *Psychological Review*, 1964, 71, 42–60.

Jacob, R. J. K., Egeth, H. E., & Bevan, W. The face as a data display. *Human Factors*, 1976, 18, 189–200.

Kahneman, D., & Tversky, A. Subjective probability: A judgment of representativeness. *Cognitive Psychology*, 1972, 3, 430–454.

Kahneman, D., & Tversky, A. Prospect theory: An analysis of decision under risk. *Econometrica*, 1979, 47, 263–291.

Lockhead, G. R. Holistic versus analytic process models: A reply. *Journal of Experimental Psychology: Human Perception and Performance*, 1979, 5, 746–755.

Lockhead, G. R., & King, M. C. Classifying integral stimuli. *Journal of Experimental Psychology: Human Perception and Performance*, 1977, 3, 436–443.

Medin, D. L., & Schaffer, M. M. Context theory of classification learning. *Psychological Review*, 1978, 85, 207–238.

Monahan J. S., & Lockhead, G. R. Identification of integral stimuli. *Journal of Experimental Psychology: General*, 1977, 106, 94–110.

Naveh-Benjamin, M., & Pachella, R. G. The effect of complexity on interpreting "Chernoff" faces. *Human Factors*, 1982, 24, 11–18.

Posner, M. I., & Keele, S. W. On the genesis of abstract ideas. *Journal of Experimental Psychology*, 1968, 77, 353–363.

Posner, M. I., & Keele, S. W. Retention of abstract ideas. *Journal of Experimental Psychology*, 1970, 83, 304–308.

Reed, S. K. Pattern recognition and categorization. *Cognitive Psychology*, 1972, 3, 382–407.

Rosch, E., Simpson, C., & Miller, R. S. Structural bases of typicality effects. *Journal of Experimental Psychology: Human Perception and Performance*, 1976, 2, 491–502.

Slovic, P., & Lichtenstein, S. Comparison of Bayesian and regression approaches to the study of information processing in judgment. *Organizational Behavior and Human Performance*, 1971, 6, 649–744.

Smith, E. E., & Medin, D. L. *Categories and concepts*. Cambridge, MA: Harvard University Press, 1981.

Summers, D. A., & Hammond, K. R. Inference behavior in multiple-cue tasks involving both linear and nonlinear relations. *Journal of Experimental Psychology*, 1966, 71, 751–757.

Wallsten, T. S. (Ed.). *Cognitive processes in choice and decision behavior*. Hillsdale, NJ: Erlbaum, 1980.

Wang, P. (Ed.). *Graphical Representation of Multivariate Data*, New York: Academic Press, 1978.

Author Index

Subject Index